Birds and Animals of Australia's Top End

Darwin, Kakadu, Katherine, and Kununurra

Nick Leseberg & Iain Campbell

PRINCETON
press.princeton.edu

Published by Princeton University Press,
41 William Street, Princeton, New Jersey 08540
In the United Kingdom: Princeton University Press, 6 Oxford Street,
Woodstock, Oxfordshire OX20 1TW
nathist.press.princeton.edu

Requests for permission to reproduce material from this work should be sent to
Permissions, Princeton University Press

First published 2015

British Library Cataloging-in-Publication Data is available

Library of Congress Control Number 2015930859
ISBN 978-0-691-16146-4

Production and design by **WILD***Guides* Ltd., Old Basing, Hampshire UK.
Printed in China

10 9 8 7 6 5 4 3 2 1

Contents

RAPTORS

BIRDS OF THE FOREST

BIRDS OF OPEN AREAS

MAMMALS

REPTILES AND AMPHIBIANS

About this book

The Top End is justly famous for its natural beauty. From towering sandstone escarpments and rugged gorges to peaceful billabongs and open savannas, visitors to this most beautiful part of Australia can experience some of the most spectacular yet serene landscapes found anywhere in Australia. There is a sense of time that exists in the Top End that is more palpable than anywhere else in Australia, and also a proximity to nature that is replicated in only a handful of other locations. This ancient landscape and the proximity to nature are just two reasons why the Top End is such a popular tourist destination. The wide variety of habitats supports an equally diverse assortment of wildlife –and perhaps more than anywhere else in Australia wildlife is an integral component of the Top End experience. After all, who is going to visit the Top End without trying to see a crocodile?

Luckily, much of the Top End's wildlife is quite easy to observe, with wildlife-watching the focus of some of the region's most popular tourist attractions. But you do not have to be on a boat or part of an organized tour to enjoy the Top End's wonderful wildlife. The independent traveller who keeps an eye out will be surprised how much wildlife can be found – from the ubiquitous Black Kites and Whistling Kites soaring over the highway, to the friendly Agile Wallabies lounging in the shade at your campground. Your enjoyment of the Top End's abundant wildlife, or wildlife anywhere, will be significantly enhanced if you know what you are looking at. Knowing what you have found opens up a new world, allowing you to delve into the fascinating natural histories of the many creatures that make the Top End their home, including where they live and how they survive. Did you know that Estuarine Crocodiles are the largest living reptile in the world, that they can live for up to 100 years, or that they may not eat for months at a time?

The primary aim of this book is to set you on the path of learning, not just by helping you identify the Top End's wildlife but also by giving you an introduction to the unique biology of these remarkable creatures. The book covers most of the birds you can expect to find, along with a selection of the mammals, reptiles and frogs you are likely to come across during your travels. For each species there is a photo, along with some information on how to identify it, how to tell it apart from any similar species, where you might find it and interesting information on that animal's natural history.

Ultimately, we hope this book is only a starting point. We can only cover a fraction of the vast amount of knowledge out there on the many wonderful creatures found across the Top End. It may be that your trip to this wonderful part of Australia kick-starts a lifelong interest in nature. Perhaps the thrill of identifying your first bird from the pages of this book leaves you wanting to do it again. If nothing else, we hope that recognizing and reading about some of the amazing creatures you might see on your travels helps you enjoy this spectacular part of Australia just that little bit more.

How to use this book

There are plenty of fantastic books available which can help you identify all of the birds, mammals, reptiles and frogs that you may come across in the Top End. However, most of these books are aimed at the already experienced wildlife enthusiast, who may visit the Top End for the sole purpose of searching out some of these obscure creatures. Their complicated text, plethora of illustrations and often large size (because they may cover the entire country), can be confusing for the more casual wildlife-watcher who may not know where to start when confronted with 900 possibilities for the bird they have just seen!

The purpose of this book is to focus both on those species you are likely to see in the Top End, and those with interesting natural history, in a way that is accessible to the novice wildlife-watcher. By doing that, it has been possible to produce a book that is small enough to carry around with you, or put in your day-pack or hand-bag, so that you can look up the animals you come across, or read about those found in the area you are visiting. Most species are illustrated with large, clear photos that will make identification easy. The text aims to be simple and informative, and to provide a summary of how to identify each species, where you might find it, its ecology – how it survives – and what makes it interesting.

Because the aim is to cover species likely to be seen by the average wildlife-watcher, there are many species that are not covered. A good example is the multitude of small reptiles you might see scurrying off the trail, or the many species of small microbats that are often seen fluttering around after dark. Identifying many of these animals requires them to be caught in order to count scales, measure wing-lengths or analyze some other obscure diagnostic feature. Getting into this level of detail can be great fun and will be the next step in your wildlife-watching pursuits. If you want to identify some of these animals, the books in the *Further reading* section (*page 264*) will help you do so. But be careful – this can become addictive!

The animals in this book are arranged into four basic groups – birds, mammals, reptiles and frogs – with each of those groups broken down further into sub-groups of similar animals. This classification of animals is a scientifically complicated process, and plenty of academics spend their entire careers trying to work out which animals fit into which groups. Generally, species that are similar to each other appear on adjacent pages, enabling you to compare them as you try to figure out what you have seen. This is a little more difficult for birds, as there are so many of them that you may have to do a bit of flicking – but, to help, the species are shown in separate sections that indicate the broad habitat types in which they are found. In addition, similar species are highlighted in the text, and, where appropriate, cross-referenced to the relevant page in the book.

To help you find as much of the Top End's wildlife as possible, we have also included a small section on some of the best wildlife-watching spots. Of course the list is not exhaustive as there is wildlife everywhere, but it may give you some ideas about where to see certain species, or may prompt you to keep an eye out for something particular as you are out and about on the tourist trail.

Each species account provides some general information about its status and distribution in the Top End, an indication of its habitat preferences, a brief description and if necessary notes on how to identify it, and, where appropriate, interesting facts about its ecology; for similar species on the same page this discussion may be combined. In most cases, guidance is given on where to find the species and, for many, specific locations to look.

Finally, there can be some confusion about the common name of some species, particularly for birds. Different organizations put out lists of common names, and sometimes cannot even agree on whether an animal is a species in its own right, or a subspecies of a different animal. This often results in some animals having a different name, depending on the organization. A good example is the well-known Bush Stone-curlew, which is often called Bush Thick-knee. Where there is an alternative name that is commonly used, this is included in the text.

Varied Lorikeet

Maps of the Top End

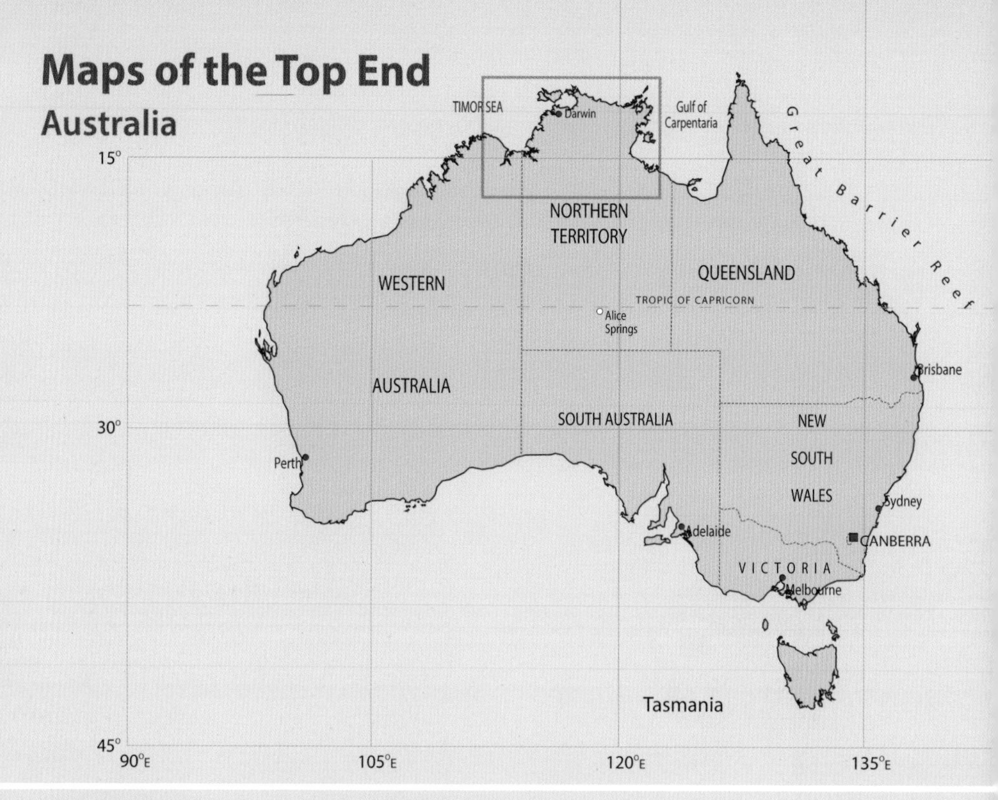

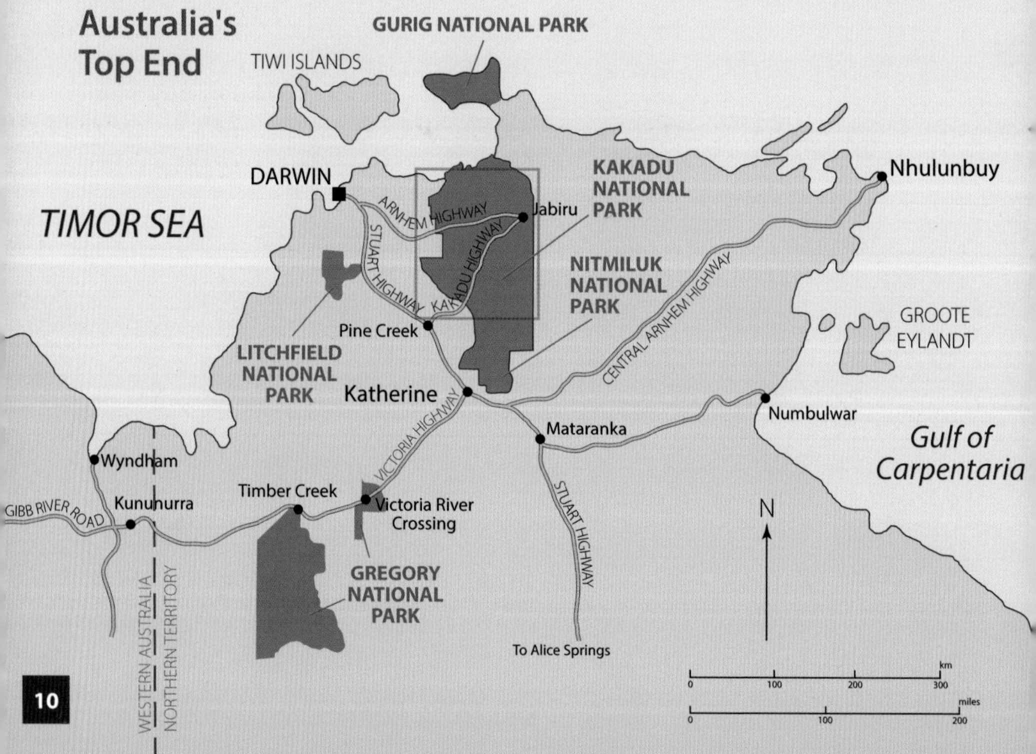

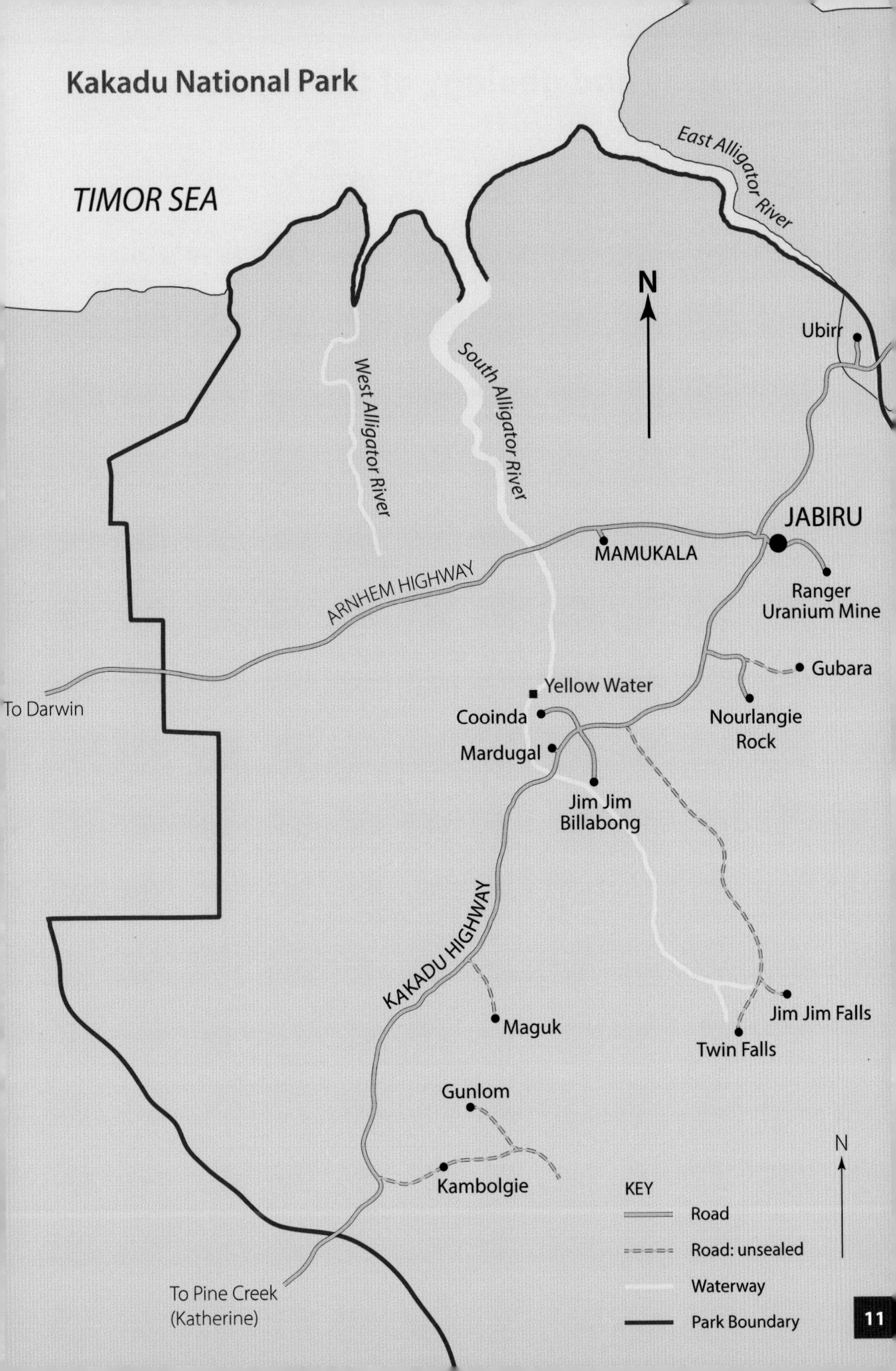

Kakadu National Park
TIMOR SEA
East Alligator River
West Alligator River
South Alligator River
N
Ubirr
JABIRU
MAMUKALA
ARNHEM HIGHWAY
Ranger Uranium Mine
Gubara
Yellow Water
To Darwin
Cooinda
Nourlangie Rock
Mardugal
Jim Jim Billabong
KAKADU HIGHWAY
Maguk
Jim Jim Falls
Twin Falls
Gunlom
Kambolgie
KEY
Road
Road: unsealed
Waterway
Park Boundary
To Pine Creek (Katherine)

Geography and geology of the Top End

The 'Top End' is a broadly defined area that covers the northern part of the Northern Territory, from the north coast south to a line extending very approximately from Timber Creek across through Daly Waters to Borroloola. Darwin, on the north-west coast, is the largest city in the region and the entry point for many visitors. About 320 kilometres (200 miles) southeast of Darwin is Katherine, the region's other main population centre. The Stuart Highway runs from Darwin to Katherine and continues south through the Northern Territory all the way to Alice Springs, and is another major route of entry to the Top End. The Victoria Highway heads south-west from Katherine through the Victoria River region, and eventually to Kununurra in Western Australia, and is a popular route for tourists particularly in the dry-season. Much of the eastern Top End is relatively inaccessible, and only visited by the most intrepid traveller, with few people venturing as far as Nhulunbuy, a small mining town in the north-east.

Geologically, much of the Top End is very old and very stable, some of the most ancient formations in the Pine Creek and Kakadu regions being more than 2 billion years old. As a result, much of the exposed rock has been severely weathered and the soils are heavily leached and lateritic, meaning that most of the nutrients such as Calcium, Potassium and Magnesium have been removed, leading to generally infertile, inert, silica-dominated and iron-dominated soils. This geological stability and constant weathering is why much of the Top End is relatively flat, with seasonal creeks, small seasonal rivers and larger tidal rivers providing the only relief. The obvious exceptions are the rugged sandstone escarpments that are scattered across the Top End; these are most prominent in Arnhem Land in the central Top End and the Victoria River region in the south-west. Sandstones and conglomerates were originally deposited in shallow seas and distal fans hundreds of millions of years ago. Since then they have undergone various periods of metamorphism, which has made them rather homogeneous and erosion resistant. The homogeneous nature of these materials has resulted in the primary erosion forces creating steep escarpments along the edges of flat-topped plateaux.

The region's coastal floodplains, particularly between Kakadu and Darwin, are an example of more recently formed landscapes. Only 15,000 years ago, during the last Ice Age, the coastline was thought to be about 320 kilometres (200 miles) farther north than it is now. As the sea level rose following the Ice Age, many of the large rivers were swamped by advancing seawater for distances of up to 100 kilometres (60 miles) inland. The subsequent deposition of sediments through wet-season run-off has slowly built up these coastal plains, lessening the influence of tidal inundation and turning them into freshwater wetlands; many of the well-known wetlands may, therefore, only be a few thousand years old.

Victoria River escarpment

Weather and seasons of the Top End

The normal northern hemisphere annual cycle of four seasons is not especially relevant in the Top End. Many residents simply refer to the 'wet' and 'dry' seasons experienced in most tropical areas, while the Aboriginal people of the Kakadu National Park region recognized six 'seasons'. Ultimately, the region's climate is driven by its position relative to the belt of high pressure encircling the globe at a latitude of 30°S – directly over continental Australia. Its position is a result of the earth's orientation relative to the sun but, importantly, the earth's rotation around the sun throughout the year causes this belt to move and creates the regular seasons and rainfall patterns seen in the Top End. The primary impact of the high pressure belt is stable, dry conditions. As this belt moves throughout the year, these conditions influence different parts of Australia – and hence the differences in climate between the tropical north and temperate south. From May to August the belt of high pressure sits over northern Australia and the Top End, resulting in the 'dry-season'. Later in the year it begins to move south, allowing the monsoonal trough that usually sits farther north near the equator to also move south, bringing warm, moist air over northern Australia and exposing the Top End to the effects of the equatorial monsoons – the wet-season, which dominates from December to March.

The dry-season (May to August) is characterized by stable conditions, with mild temperatures across the Top End and very little, if any, rain. Average daily maximum temperatures are around 30–32°C (85–90°F) in Darwin and similar in Katherine. Average minimum temperatures are around 19–21°C (66–70°F) in Darwin, and cooler around Katherine at 13–15°C (55–60°F). Humidity is generally low, making it a very pleasant time of year. The transition from the 'dry' to the 'wet' season occurs between September and December, and is often referred to as the 'build-up'. It is the hottest time of year, and although temperatures in Darwin increase by only a couple of degrees, temperatures farther inland increase significantly, with average daily maxima in Katherine around 37–39°C (98–102°F). The humidity at this time of year increases significantly, making conditions quite oppressive. It occasionally rains during the 'build-up', but usually as isolated cells, bringing only temporary respite from the hot conditions. The wet-season proper arrives in January and is dominated by monsoonal depressions, heavy rain and occasional tropical cyclones that can bring torrential rain and strong, destructive winds. Around 60–70% of the annual rainfall occurs during January, February and March, with 90% of the annual rainfall occurring between November and April. Average daily temperatures decrease slightly during the wet-season, and humidity is constantly high. The wet-season generally begins to wind down around March, transitioning back to dry-season conditions by May.

These huge variations in weather conditions have a marked impact on the Top End's environment, and have a significant impact on where you can visit and what you can see at any particular time. The wet-season is a difficult time to visit, with seasonal waterways swollen, many roads closed due to flooding and some of the primary tourist attractions such as Katherine Gorge and parts of Kakadu National Park inaccessible. It can be difficult to comprehend the volumes of water that fall during this period. In 1998 the Katherine

River flooded the town, with the river peaking at 20·4 metres deep. Look at the markers on the bridge and imagine that volume of water as you head north out of Katherine! As the wet-season comes to a close and the floodwaters start to recede, roads begin to re-open, although if it has been a 'big' wet-season, it can be some months before all roads are open again. In response to the rain, the grassy savannas spring to life and the landscape is lush and green. This can be a spectacular time to visit the Top End as most of the rivers are still flowing strongly, particularly the waterfalls that cascade off the sandstone escarpments of Arnhem Land and in other places such as Litchfield National Park. These conditions continue into the early dry-season as the weather becomes pleasant and sunny, making it one of the most popular times to visit the Top End.

As the dry-season progresses conditions slowly change. The lush, green grass begins to dry out, providing ample fuel for the small fires that spring up across the landscape. The seasonal watercourses continue shrinking and much of the standing water begins to dry up. The weather, though, is at its best, and this is a great time to visit the Top End. As the dry-season draws to a close and the 'build-up' begins, temperatures increase and water becomes scarce in the landscape. It can be an uncomfortable time to visit the Top End, but is the best time for wildlife-watching. After months without rain, most seasonal watercourses are now barely a trickle and there is very little standing water around. Wildlife is easier to see as it congregates around the water that does remain, making this the best time for activities such as a boat trip on Kakadu National Park's Yellow Water billabong, or searching for birds around Timber Creek. As the 'build-up' continues and rain begins to fall, the landscape comes to life again, and the cycle is completed.

Oncoming storm

Habitats of the Top End

The patterns of prevailing climate, water availability, soil type and local topography, as well as a number of other factors, combine to determine the vegetation that grows in any particular area. The physical environment and type of vegetation present are collectively referred to as the habitat and since this influences the animals that may occur in that area, it is helpful for the wildlife-watcher to have an understanding of what the basic habitat types are. The classification of habitats and vegetation types can be very complex, but this section of the book provides a simple breakdown of the main habitats found in the Top End, which should help you to predict where certain animals are likely to occur.

Beaches and mudflats

Beaches and mudflats are primarily coastal habitats which are heavily influenced by the tides, being inundated and exposed daily. This provides rich feeding grounds for the large number of migratory shorebirds that spend the summer in Australia. This group of birds includes some of the world's most incredible travellers, migrating annually between their breeding grounds in the far northern hemisphere and their wintering grounds in Australia – a round trip of more than 25,000 kilometres (15,000 miles)! There are plenty of accessible examples of this habitat close to Darwin, including Buffalo Creek, Lee Point and Nightcliff.

Mangroves

Mangroves are very special trees that are adapted to grow in a saline or saltwater environment. They are usually found in the intertidal zone, such as on beaches and inlets, and along tidal rivers, where they form an important part of the marine ecosystem. Access to mangroves can be difficult and sometimes dangerous, with thick mud, crocodiles and biting insects all hazards that can make them unpleasant to visit. A suite of bird species is restricted to mangroves, including Chestnut Rail, Mangrove Golden Whistler and Mangrove Robin, and this habitat type is therefore of particular interest to birdwatchers. Luckily, there are several places where mangroves are easily accessible, including Buffalo Creek and Nightcliff near Darwin.

Rivers and billabongs

Large rivers like the Roper, South Alligator and Victoria Rivers are an integral part of the Top End's landscape. Winding their way through sandstone gorges and across floodplains, they provide habitat for some of the region's iconic wildlife, most famously the Estuarine Crocodile. These rivers are often tidal and carry water year-round. Smaller watercourses further inland like the Waterhouse River near Mataranka, and Edith River near Katherine may not be as large, but they still always carry some water, water that the wildlife of the Top End relies on to survive the dry-season. True billabongs are sections of river isolated after the river changes course, much like an oxbow lake, but the term is often used to refer to any waterhole or isolated section of a small creek or river that may hold water. These billabongs can be magnets for wildlife and are therefore great places for wildlife-watching, particularly if they retain water late into the dry-season, as myriad birds and mammals flock to them to drink, particularly in the late afternoon or early morning.

Freshwater wetlands and floodplains

Some of the Top End's most popular wildlife-watching destinations are its freshwater wetlands and floodplains. Many of the wetlands in Kakadu NP are seasonally flooded river plains, inundated during the wet-season and slowly shrinking during the dry, while in other areas natural depressions that are refilled every wet-season form freshwater wetlands. When these wetlands or floodplains are deep enough to hold water year-round, the late dry-season can be a spectacular time to visit. As smaller wetlands dry out, ducks, geese and other waterbirds are forced to find other water sources, until clouds of waterbirds, plus other wildlife such as crocodiles, turtles and fish are concentrated around the few wetlands that remain. Kakadu NP has some of the most accessible freshwater wetlands in the Top End, including Mamukala and the magnificent Yellow Water billabong, while closer to Darwin permanent waterholes such as Fogg Dam and Knuckey Lagoons can be good places to see wildlife in the late dry-season.

Monsoon forest

Monsoon forest, also known as dry rainforest (*left*), is not a widespread habitat in the Top End but is home to some of the region's unique wildlife. Only found in small, isolated patches, monsoon forests are quite different from the tall, lush rainforests of eastern Australia, which occur on more fertile soils. The rainforests in the Top End are generally restricted to river margins or other areas which receive a lot of water, such as the bases of sandstone escarpments. They are of particular interest to wildlife-watchers, as they are home to such spectacular birds as the Rainbow Pitta and Orange-footed Scrubfowl. Close to Darwin there are good examples of monsoon forests at East Point Reserve, while the shady forests around the rock art galleries at Nourlangie Rock in Kakadu NP are another good example.

Sandstone escarpments

Sandstone escarpments (*right, below*) are one of the most spectacular habitats in the Top End, and the focus of many of the region's most popular tourist attractions. From sheer cliffs to jumbled fields of boulders, there are examples scattered across the region, with the most famous being the Arnhem Land escarpments of Kakadu NP. Litchfield NP, Katherine Gorge and Gregory NP are other areas of impressive escarpment country. Exploring this rocky habitat can be physically a little demanding, but is worthwhile as it is home to many species found nowhere else, including a variety of geckos, the Oenpelli Python, rock-wallabies, and several unique birds such as the Black-banded Fruit-dove and White-throated Grasswren. Nourlangie Rock in Kakadu NP is an easily accessible sandstone escarpment, and there is a fantastic escarpment walk near Victoria River Crossing. For the intrepid traveller, the Jatbula Trail from Katherine Gorge to Edith Falls is a multi-day hike along the southwestern edge of the Arnhem Land escarpment.

Savanna woodlands

The most widespread habitat in the Top End, sometimes the savanna woodlands seem to stretch on forever. Characterized by scattered eucalyptus trees and a grassy understorey, they change significantly as the seasons progress, supporting lush green grass after the wet-season, which slowly dries out and is often burnt before the end of the dry-season, revealing the red earth beneath. There is some variation across the region, with different species of eucalypt or grass dominating. The savanna woodlands are home to some of the region's most sought-after birds, including the gorgeous Gouldian Finch and Hooded Parrot, while grazing mammals like the Agile Wallaby and Antilopine Wallaroo are also common. Since this is the Top End's dominant habitat type, it is easily accessible at many locations.

How to watch wildlife

The fantastic thing about wildlife-watching is that we can all do it, wherever we are. From Darwin's city centre to the most remote parts of the Northern Territory, there is wildlife everywhere, waiting to be seen. Wildlife can be seen by simply keeping an eye out while you are travelling down the highway, or by taking a short walk, although sometimes an intense search is needed to find a particular species. This section outlines a few simple techniques, and suggests some equipment, that will improve your chances of both finding and identifying not just the Top End's wildlife, but wildlife anywhere.

Equipment

The one piece of equipment that will most improve your wildlife-watching experience is a pair of binoculars. Even a small and inexpensive pair will bring the wildlife much closer, not only allowing you to pick out features and see much more detail than you might see with the naked eye, but also to observe an animal's behaviour more clearly. A small camera can also be handy, not only for recording what animals you have seen, but also for looking back at animals you may have seen but could not identify at the time. Most small pocket cameras, and even many smart-phones, have quite capable zoom functions that will enable you to take a reasonable photo to check later on.

When to watch wildlife

The weather in the Top End is often quite warm and, just like us, most animals prefer to be active during the cooler parts of the day. This means that early morning and late afternoon are often the best times to see diurnal species, as they move about to feed or find water. Particularly late in the dry-season, when there is little standing water around, watching a waterhole late in the afternoon can be a very enjoyable experience as a steady stream of finches, parrots, honeyeaters and wallabies come to drink.

Star Finches

Billabong at dusk

Finding wildlife

Torresian Imperial-pigeon

Finding wildlife takes practice; the more often you get out looking for birds and animals, the better at it you will become. One of the keys to finding some of these creatures is knowing a little about their habits, giving you an idea of the best places to look. Obviously all animals need to eat and drink, so fruiting or flowering trees are likely to attract hungry lorikeets, honeyeaters and flying-foxes, while waterholes will be good places to look for animals that need water, such as wallabies, finches and parrots. Each species has slightly different requirements, but as you learn about some of the Top End's amazing wildlife and how they survive in their environment, you will start to get an idea of where you might find each species, and how best to look for it.

Watching nocturnal species

Many of the Top End's mammals, reptiles and frogs do not emerge from their hiding places until after dark, so to find them you also need to head out after dark. A small torch can suffice, but you will find more with a larger torch or spotlight. Just wandering around a campsite at night may turn up possums, flying-foxes or wallabies, while after rainy periods you will be surprised at the number of frogs that materialize, hopping around everywhere. The best way to find animals with your spotlight is to hold it as close to your eyes as possible and look for the animal's eye-shine being reflected back at you. This can take a bit of practice, but you will soon get the hang of it.

Where to find wildlife

You can watch wildlife practically anywhere in the Top End. Rufous Owls have even been seen hunting flying-foxes along Mitchell Street in the Darwin Central Business District with crowds of late-night revellers partying on the street below. Although a visit to Mitchell Street at night would not be recommended as a good place to start your wildlife-watching adventure, the point is you can do it almost anywhere! Of course some places are better than others for wildlife-watching, and here is a selection of some of the best sites to enjoy the Top End's wonderful wildlife. More information on all of these sites can be found on the Northern Territory's Parks and Wildlife Commission website www.parksandwildlife.nt.gov.au.

Fogg Dam Conservation Reserve

Located about an hour's drive from Darwin along the Arnhem Highway, Fogg Dam is a large wetland which can harbour thousands of waterbirds, including Magpie Geese and Black-necked Storks, along with Estuarine Crocodiles and Water Pythons. There are also patches of monsoon forest and open woodland. The park is very accessible with a number of trails and viewing platforms, along with guided walks by local rangers during the dry-season.

Gregory National Park

This large national park in the southwest of the Top End protects extensive areas of semi-arid landscape and rugged sandstone escarpments, quite different from areas farther to the north and east. Victoria River Roadhouse provides access to the Victoria River where the lovely Purple-crowned Fairywren can be found. Agile Wallabies are common around the campground, while the nearby escarpment walk is a great place to see Short-eared Rock-wallabies. Timber Creek is a birdwatching mecca, and one of the most reliable sites for that jewel of the Top End, the Gouldian Finch. The small watercourse through Timber Creek itself is a great place to see Freshwater Crocodiles.

Yellow Water

Kakadu National Park

Kakadu NP is not only one of the best places in Australia to watch wildlife, but one of the best places in the world. The premier wildlife-watching experience in Kakadu is the fantastic Yellow Water cruise, where clouds of waterbirds, hulking Water Buffalo and stealthy Estuarine Crocodiles can be seen almost within touching distance. Mamukala is another great wetland for birds, while Nourlangie Rock and the escarpment climb at Gunlom allow access to the sandstone escarpments and their unique wildlife.

Howard Springs Nature Park

Located about 25 minutes' drive south of Darwin, Howard Springs is a great place to enjoy the natural beauty of the Top End. There is a large waterhole surrounded by monsoon forest with an easy walking trail. In the forest you can see Orange-footed Scrubfowl, Rufous Owl, Rainbow Pitta and Rose-crowned Fruit-dove if you are patient, and there is a large Black Flying-fox colony near the waterhole. By standing quietly on the bridge over the springs you will often turn up a few Northern Yellow-faced Turtles.

Mataranka and Elsey National Park

An hour's drive south of Katherine, the primary attraction at Mataranka is Elsey National Park, where the Waterhouse and Roper Rivers provide swimming opportunities for tourists and habitat for a number of animals. Agile Wallabies are common in the open woodlands, and the larger Antilopine Wallaroo can sometimes be seen. There are plenty of birds around, with Blue-winged Kookaburras common, along with gaudy parrots like Red-collared Lorikeets and Red-winged Parrots. It is also a hotspot for the elusive Red Goshawk, one of Australia's rarest birds.

Nitmiluk National Park and Katherine Gorge

Most famous for its natural beauty, Katherine Gorge is also a great place to keep an eye out for wildlife. Agile Wallabies are sometimes seen around the park entrance, while a huge flying-fox 'camp' can be found in the gallery vegetation by the river. The gorge itself is home to Freshwater and sometimes Estuarine Crocodiles, while Great Bowerbirds and Blue-faced Honeyeaters are easy to see around the park headquarters.

Katherine Gorge

Birds of Wetlands and Beaches

This section covers all those species that spend much of their lives in and around water. From sandy beaches to rocky shores and freshwater wetlands, many birds spend their entire lives near water, and they are often well-adapted for this lifestyle. Many have long legs or long bills to help them wade in water or probe the mud for food, while others have webbed feed that help them swim. Some, like the two small kingfishers, feed exclusively on fish or other creatures found only in water, so are always found near water.
The Top End contains some of Australia's most spectacular wetlands, with Yellow Water in Kakadu National Park perhaps the most famous, making it a great place to see many of these fantastic birds.

Pied Heron

Comb-crested Jacana

Silver Gull
Little Kingfisher
Black-fronted Dotterel
Plumed Whistling-duck

Where to find Possible on almost any wetland across the Top End. The best place to see these birds is Yellow Water in Kakadu NP although they can also be seen around Darwin itself at the Knuckey Lagoons, and at Fogg Dam.

☐ Magpie Goose *Anseranas semipalmata* 27½–39 in | 70–99 cm

These large geese can be found on wetlands anywhere across the Top End, but are most common around Darwin and on the floodplains of Kakadu NP. They are very social, usually seen in groups from just a few birds, up to flocks of thousands, and are most often found grazing on grassy areas around the margins of billabongs and lagoons. Magpie Geese can also be seen swimming and foraging among floating vegetation, sometimes upending and using their hooked bills to collect Water Chestnut bulbs from the mud below the water surface. They were a popular food source for Aborigines, and the managed harvest of some eggs and also adult birds is still permitted. As the dry-season progresses and wetlands dry up, Magpie Geese seek refuge on remaining lagoons and this is when large congregations are most likely to occur. After the wet-season the geese breed, usually in family groups of three birds, with a male and two females building a nest and sharing the duties of incubating the eggs and raising the chicks. Easily recognised, these large black-and-white birds are often seen in pairs or threes, and you can usually identify the male by his larger size and a larger knob on top of his head.

☐ Plumed Whistling-duck *Dendrocygna eytoni*

16–23½ in | 40–60 cm

Plumed Whistling-ducks tend to be more common than Wandering Whistling-ducks, and also more likely to be seen in large flocks. They are quite noisy, and as you get close to a group you will hear them start twittering and chattering excitedly. They move around searching for wetlands, often at night when you may hear their thin whistles – "*wee-whew, wee-whew*" – as they pass overhead in the dark. Plumed Whistling-ducks are pale brown, with darker brown wings and a white throat. The sides of the breast are chestnut, and they have distinctive tufts of long, stiff, pale cream feathers that grow from their flanks. A good feature to separate them from the similar Wandering Whistling-duck is the colour of the bill and legs: those of Plumed Whistling-ducks are pink, whereas in Wandering Whistling-ducks they are black.

Where to find Yellow Water in Kakadu NP is a good place to see this species, but it can also be found at Knuckey Lagoons in Darwin, and at most other wetlands.

☐ Wandering Whistling-duck *Dendrocygna arcuata* 19½–23½ in | 50–60 cm

Wandering Whistling-ducks are most often found in small groups of twos or threes scattered across a wetland, but if you find a large flock of Plumed Whistling-ducks, examine it carefully as you will often find a smaller group of Wandering Whistling-ducks mixed in, or sitting at the edge of the flock. Quite similar to Plumed Whistling-duck, Wandering Whistling-ducks have a more chestnut-coloured body and neck, with a black crown and back of the neck, and very dark brown wings. They also have cream-coloured flank plumes, but these are much shorter and less obvious. A good way of separating the two species is to check the colour of the bill and legs: they are black in Wandering Whistling-duck but pink in Plumed Whistling-duck.

Where to find Yellow Water in Kakadu NP is the best place to see this species. It can also be found at most other large wetlands.

☐ Radjah Shelduck *Tadorna radjah* 19½–23½ in | 50–60 cm

This striking duck is quite common across the Top End and can turn up on any body of water, particularly in the early dry-season when pairs may be found even on the smallest pool. Unlike most other ducks, it can tolerate brackish water, and is sometimes seen resting on sand bars at the beach or flying along mangrove-lined rivers. Radjah Shelduck is very easy to identify, having a white head, neck and body, dark chestnut wings and a distinctive chestnut breast-band.

Where to find Possible on most freshwater wetlands and occasionally near the beach. The best place to see this species is Yellow Water in Kakadu NP.

Australasian Grebe

Tachybaptus novaehollandiae
9–10 in | 23–25 cm

Most permanent wetlands across the Top End have a few resident pairs of this grebe, and they can sometimes even turn up on very small wetlands such as farm ponds. They are nearly always seen swimming, like a little rubber duck, on open water or among floating vegetation, and spend much of their time diving for food below the surface. When breeding, usually later in the year, they develop a chestnut patch on the side of the neck and a small yellow spot near the bill.

Where to find Good places to look are Yellow Water in Kakadu NP, Fogg Dam, and Knuckey Lagoons in Darwin.

Green Pygmy-goose

Nettapus pulchellus
12½–15 in | 32–38 cm

These lovely little geese can be found on most permanent wetlands across the Top End, where they are usually seen in pairs swimming among floating vegetation such as water-lilies. Pygmy-geese have glossy green wings and fine scalloping on the breast and belly. Males have a glossy green neck and head with a white cheek patch, while females have a finely barred neck, white cheek patch and dark crown.

Where to find Can be found on most large permanent wetlands including Yellow Water in Kakadu NP, Fogg Dam, and Knuckey Lagoons in Darwin.

Hardhead *Aythya australis* 17–19½ in | 43–50 cm
(White-eyed Duck)

A nomadic species that may occur on any wetland at any time, it is more often seen on large wetlands with open water such as dams than other waterfowl. Like Grey Teal and Pacific Black Duck, this species is most common in the region during the dry-season, although some birds are present year-round. Unlike most Australian ducks which feed at the surface of the water, the Hardhead often dives, feeding on aquatic vegetation and invertebrates in the muddy substrate. The body and head are deep-chestnut in colour, with white under the tail. The underwings and belly are white but this is only seen when the bird flies. Males have distinctive white eyes, hence the alternative common name.

Where to find Can occur almost anywhere there is open water.

Pacific Black Duck *Anas superciliosa* 18½–23½ in | 47–60 cm

This is one of Australia's most common ducks, and the familiar duck of urban pools and ponds, where birds often become accustomed to being fed. Most wetlands throughout Australia hold at least a few pairs, and although not as common in the Top End as in the eastern states, it may still be found on any wetland. It is most often seen in pairs or small groups, but rarely gathers in large flocks like other species of duck. Although present year-round, it is most common during the dry-season, with some birds moving out of the region during the wet-season, presumably migrating to southern Australia. Despite its name, this duck is actually brown and has a buff face and throat, with a dark brown cap, and a dark brown line through the eye.

Where to find Often seen at smaller wetlands such as the Knuckey Lagoons in Darwin, but seems to avoid larger wetlands such as Yellow Water in Kakadu NP.

Grey Teal *Anas gracilis* 16½–17½ in | 42–44 cm

This nomadic species may turn up on any wetland across the Top End, but is most often found during the dry-season. A few birds sometimes remain during the wet-season, but most leave the region around November before returning around the middle of the year. Usually found in pairs or small groups, this species can be seen loafing on the edge of a lagoon, or swimming close to the shore, feeding on aquatic vegetation and invertebrates they collect from on or just below the surface. It is a small duck and males and females are identical. The body and head is brownish-grey, with feathers on the breast edged pale brown, giving a scalloped appearance. The face and throat are silvery-grey.

Where to find Can be seen on almost any wetland, but its presence is difficult to predict.

Hardhead
Pacific Black Duck
Grey Teal

☐ Black-necked Stork *Ephippiorhynchus asiaticus* 43½–53 in | 110–135 cm
(Jabiru)

These stately birds are one of the Top End's iconic species, and seeing them is always a highlight of any trip to the region. They are found right across the Top End, usually on larger lagoons, billabongs and wetlands, but they do sometimes turn up on smaller, isolated ponds or streams. They are commonly called 'Jabiru', a name which is mistakenly assumed to be Aboriginal, but is actually of Brazilian origin. 'Jabiru' is the name of a very similar bird that lives in South America, and which was somehow appropriated for this species, presumably by early explorers who had seen the South American species. Many large permanent wetlands have a resident pair of Black-necked Storks, and they are often seen wading around in the shallows, or on open plains near water, searching for fish, frogs, reptiles and other prey. Their nest is a large platform of sticks in the top of a tall tree. These huge birds are unmistakable, standing more than 4 feet tall, on long, bright red legs. The long neck, head and huge bill are glossy black, with a bluish sheen if seen in good light. Younger birds are dull brown instead of black. It is possible to tell the males and females apart: female birds have a yellow eye, whereas males have a dark eye.

Where to find Although they can turn up on almost any wetland across the Top End, Yellow Water in Kakadu NP is one of the best places to see this species, and there is usually a pair at Knuckey Lagoons in Darwin. They can also be seen at Fogg Dam.

☐ Australian Pied Cormorant

Phalacrocorax varius

25½–33½ in | 65–85 cm

Usually seen sitting on rocks by the coast or on trees by large rivers, this species is the least common cormorant in the Top End. It is much larger than Little Pied Cormorant, and similar in being black above and white below, but has a longer, more slender bill. If seen well, you will notice a small patch of yellow skin between the bill and the blue eye-ring.

Where to find Can be found around the coast near Darwin, or along any of the large estuaries or rivers.

☐ Little Pied Cormorant

Microcarbo melanoleucos

21½–24 in | 55–61 cm

Common right across the Top End, these little cormorants can be found on almost any body of water, however large or small, from the coast, to small billabongs or farm ponds. They are often seen sitting in a tree beside the water, sometimes holding their wings out to dry. Like the Australian Pied Cormorant, they are black above and white below but have a relatively small, yellow bill.

Where to find Easily seen at Yellow Water in Kakadu NP and at many other wetlands such as Fogg Dam.

Little Pied Cormorant

Australian Pied Cormorant

□ **Little Black Cormorant** *Phalacrocorax sulcirostris* 21–25½ in | 53–65 cm

These birds are usually seen swimming low in the water, diving below the surface to catch small fish. They also perch on waterside snags, holding their wings out to dry after a bout of fishing. Unlike other cormorants, they regularly form small flocks and may fish cooperatively, moving across the water as a group, continually diving and herding fish before them. If you see them well, you should be able to make out the jade-green eye.

Where to find Found on water bodies throughout the Top End, from the ocean to small farm ponds.

Little Black Cormorant

□ **Great Cormorant** *Phalacrocorax carbo*

31½–33½ in | 80–85 cm

In the Top End, this bird is much less common than Little Black and Little Pied Cormorants, and is usually seen singly or in small groups around larger bodies of water, from the coast inland to large rivers and dams. Similar to the smaller Little Black Cormorant but much larger, it has a yellow patch of skin around the base of the bill, and usually a pale throat.

Where to find Possible on any large body of water from Yellow Water in Kakadu NP to the coast.

Great Cormorant

Where to find Good places to see this bird include Yellow Water in Kakadu NP, Fogg Dam, and Copperfield Dam near Pine Creek.

☐ Australasian Darter *Anhinga novaehollandiae* 34–37 in | 86–94 cm

Common across the Top End, these birds prefer large, usually shallow wetlands. They are most often seen swimming on the surface and repeatedly diving for fish, or sitting on a snag by the water, holding their wings out to dry after a bout of fishing. When swimming, they sit very low in the water, with their body completely submerged, showing just a long, skinny neck above the surface, usually with the bill pointing upwards. They are well equipped for their essentially aquatic lifestyle, with large, webbed feet that they use to propel themselves underwater as they search for prey. Skilful hunters, they bring any fish they catch to the surface, flipping it around in their bill so it can be swallowed head-first – a process that can be quite amusing to watch, particularly if it is a large fish. Large and bulky with a long, thin neck, small head and a dagger-like yellow bill, males are black with white streaks on the wings, a chestnut patch on the neck, and a white stripe on the side of the head. Females are similar but browner, and whitish below. Darters are often seen soaring on thermals overhead, with their long neck, broad wings and long square-tipped tail giving them a distinctive triangular silhouette.

☐ **Australian Pelican** *Pelecanus conspicillatus*

63–71 in | 160–180 cm

This iconic and unmistakable Australian species is usually common across the Top End, where it is found around the coast, and also on larger wetlands, billabongs and rivers. The lower half of the huge pink bill can expand, and the bird uses this when feeding, lunging forward and plunging the bill into the water, scooping up a huge volume of water containing, hopefully, a few fish. Birds sometimes fish communally, working together to herd fish into shallow water where they are easier to catch. Pelicans become accustomed to being fed, and will often gather around fish-cleaning stations, gobbling up scraps and competing aggressively for the loot with other pelicans and Silver Gulls (*page 72*).

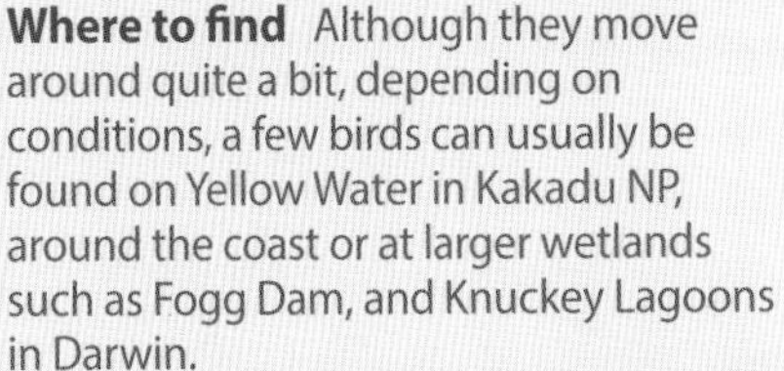

Where to find Although they move around quite a bit, depending on conditions, a few birds can usually be found on Yellow Water in Kakadu NP, around the coast or at larger wetlands such as Fogg Dam, and Knuckey Lagoons in Darwin.

Where to find Good places to look are Yellow Water in Kakadu NP and Knuckey Lagoons in Darwin.

☐ Plumed Egret *Ardea plumifera*

(Intermediate Egret)
21½–27½ in | 55–70 cm

Common on freshwater wetlands and billabongs across the Top End, this species, like Great Egret, is usually seen foraging in the shallows. It is smaller and stockier than the Great Egret and has a stouter neck that is about the same length as the body. The bill is yellow, but may be orange if the bird is breeding, when it will also have a tuft of long, thin plumes on the back and breast.

☐ Cattle Egret *Bubulcus ibis*

19½–21½ in | 50–55 cm

A relatively recent arrival to Australia, this species was not recorded regularly until the 1950s. Since then it has spread across the continent, and is now quite common in the Top End. As the name suggests, it is almost always found around cattle, following them in paddocks or congregating around cattle yards. The Cattle Egret is quite stocky for an egret, similar to Plumed Egret but smaller. In breeding plumage it develops orange or buffy plumes on the head, breast and back.

Plumed Egret

Cattle Egret Breeding

Where to find Always around or near cattle, and quite common on the outskirts of Darwin.

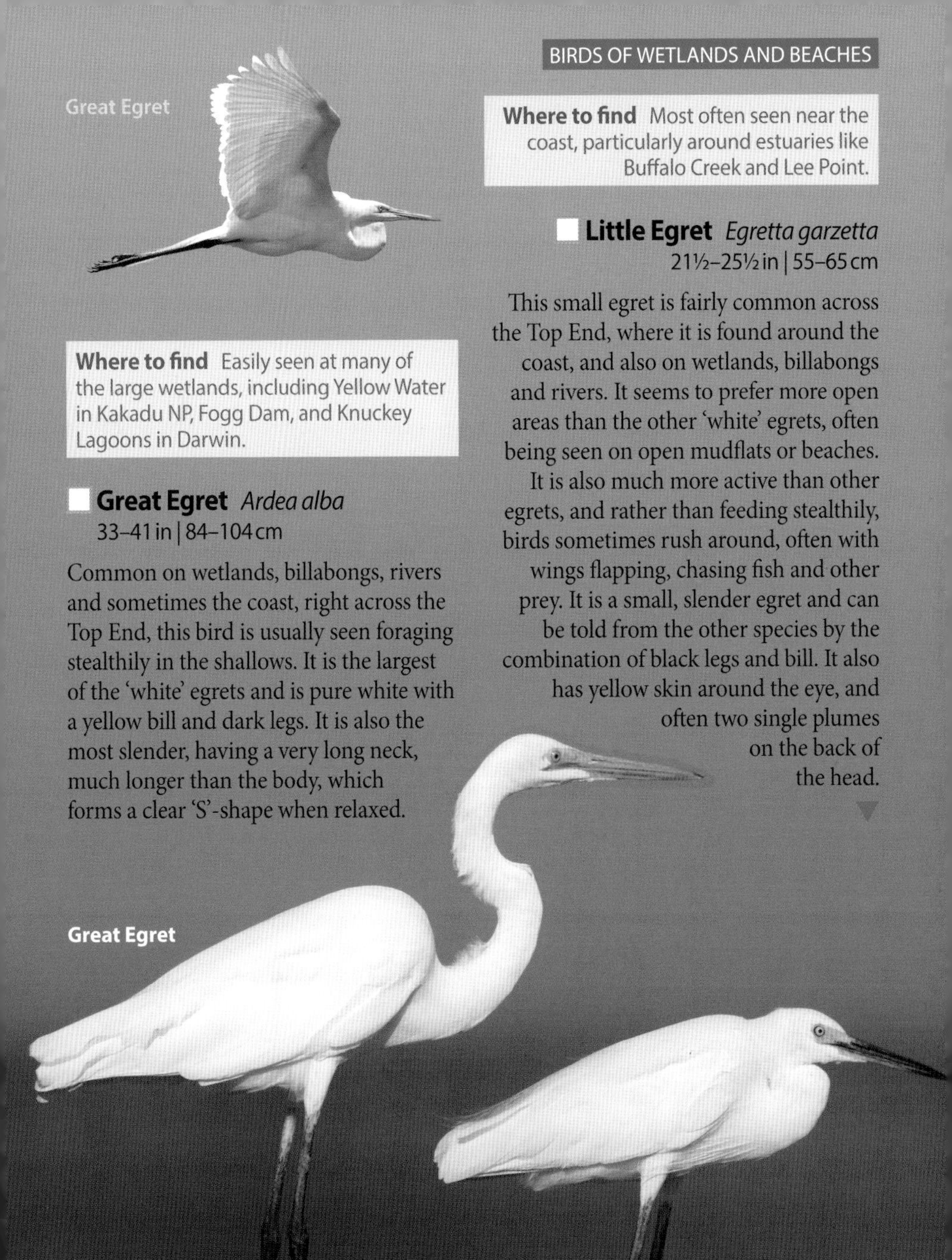

Great Egret

Where to find Easily seen at many of the large wetlands, including Yellow Water in Kakadu NP, Fogg Dam, and Knuckey Lagoons in Darwin.

Great Egret *Ardea alba*

33–41 in | 84–104 cm

Common on wetlands, billabongs, rivers and sometimes the coast, right across the Top End, this bird is usually seen foraging stealthily in the shallows. It is the largest of the 'white' egrets and is pure white with a yellow bill and dark legs. It is also the most slender, having a very long neck, much longer than the body, which forms a clear 'S'-shape when relaxed.

Where to find Most often seen near the coast, particularly around estuaries like Buffalo Creek and Lee Point.

Little Egret *Egretta garzetta*

21½–25½ in | 55–65 cm

This small egret is fairly common across the Top End, where it is found around the coast, and also on wetlands, billabongs and rivers. It seems to prefer more open areas than the other 'white' egrets, often being seen on open mudflats or beaches. It is also much more active than other egrets, and rather than feeding stealthily, birds sometimes rush around, often with wings flapping, chasing fish and other prey. It is a small, slender egret and can be told from the other species by the combination of black legs and bill. It also has yellow skin around the eye, and often two single plumes on the back of the head.

Great Egret

Little Egret

Great-billed Heron

Ardea sumatrana

39½–43½ in | 100–110 cm

This bird is famously shy and very difficult to see well. It is found along large rivers, both brackish and freshwater, usually in areas with thick vegetation such as mangroves or gallery forest, where it keeps well hidden. It is a large, grey heron with a long, thick neck, large black dagger-like bill and relatively short legs.

Where to find Cruising slowly in a small boat along secluded creeks or rivers is the best way to find this heron; it is occasionally seen on Yellow Water in Kakadu NP, but is a possibility along any of the larger tidal rivers in the Top End.

White-faced Heron

Egretta novaehollandiae

26–26½ in | 66–67 cm

Common throughout Australia, this heron may turn up wherever there is water, fresh or brackish, along the coast and rivers, and around wetlands, whether large or small. It is often seen stalking gracefully along the water's edge, but is just as likely to be seen feeding on open ground such as urban parks or playing fields. A small, slim heron, it has a grey body and neck, and a white mask around the face.

Where to find Possible almost anywhere that there is water or open ground.

Great-billed Heron

White-faced Heron

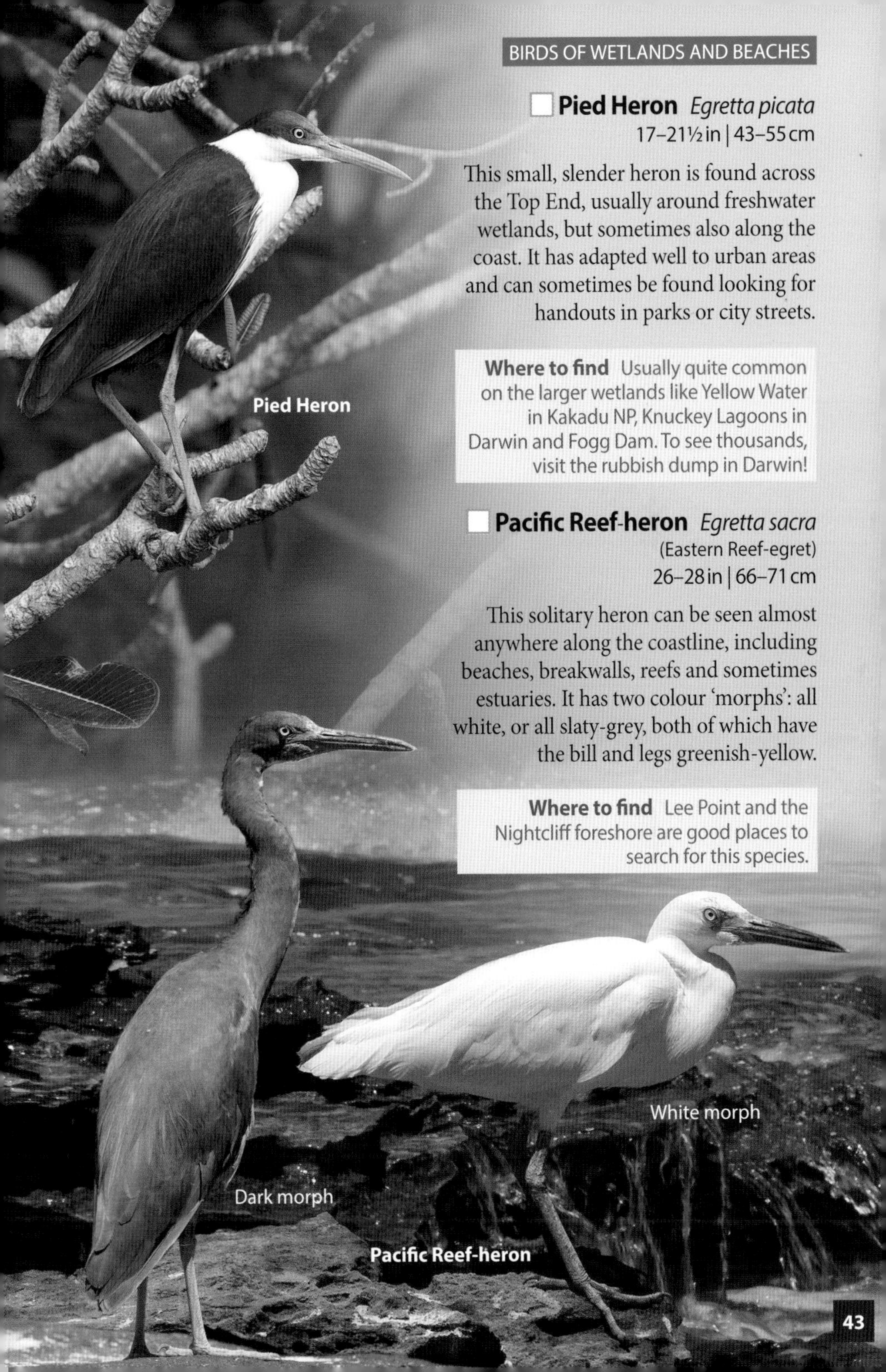

☐ **Pied Heron** *Egretta picata*

17–21½ in | 43–55 cm

This small, slender heron is found across the Top End, usually around freshwater wetlands, but sometimes also along the coast. It has adapted well to urban areas and can sometimes be found looking for handouts in parks or city streets.

Where to find Usually quite common on the larger wetlands like Yellow Water in Kakadu NP, Knuckey Lagoons in Darwin and Fogg Dam. To see thousands, visit the rubbish dump in Darwin!

☐ **Pacific Reef-heron** *Egretta sacra*

(Eastern Reef-egret)

26–28 in | 66–71 cm

This solitary heron can be seen almost anywhere along the coastline, including beaches, breakwalls, reefs and sometimes estuaries. It has two colour 'morphs': all white, or all slaty-grey, both of which have the bill and legs greenish-yellow.

Where to find Lee Point and the Nightcliff foreshore are good places to search for this species.

Where to find There are usually a few to be seen on the Yellow Water cruise in Kakadu NP, but they can turn up on any watercourse across the Top End.

☐ Rufous Night-heron *Nycticorax caledonicus* 21½–25½ in | 55–65 cm
(Nankeen Night-heron)

Most Australians call this bird 'Nankeen' Night-heron, a reference to its rich, reddish colouring which is similar to the colour of nankeen, a type of orangey cotton fabric historically manufactured and exported from Nanjing (Nanking) in China. Some of the indigenous names for this bird refer to its apparent vanity – since it spends much of its time at the water's edge staring at its own reflection. Of course it is really searching for fish, but it is a good story! It is a shy species and, as the name suggests, primarily nocturnal, although it does sometimes forage during the day. Birds spend most of the day sheltering in dense waterside vegetation, sometimes in small groups, emerging at dusk to feed around the water's edge. A stocky heron, this species has an unusual hunched, 'neckless' appearance.

Striated Heron *Butorides striata*

(Green-backed Heron)

17 in | 43 cm

This unobtrusive heron is actually quite common across the Top End, although it can be shy and difficult to see. It is usually found stalking along the muddy edges of rivers and creeks, or flying low across the water. Small and stocky, it is slaty-grey above and either grey or brownish below, with a black cap and yellow legs. Young birds are browner, and more streaked below.

Where to find Good places to search include Buffalo Creek near Darwin, or any accessible areas of mudflats and mangroves.

Black Bittern

Ixobrychus flavicollis

22–26 in | 56–66 cm

This very shy bird is not common and can be difficult to see, as it spends most of its time under the cover of dense stream-side vegetation. A fairly small, stocky heron, adults are black above with a yellowish throat and black and white streaks on the breast. Young birds are browner and more streaked, looking quite different from the adults.

Where to find Sometimes seen on Yellow Water in Kakadu NP, but may occur on any thickly vegetated watercourse from the coast to Katherine.

Australian (White) Ibis

Threskiornis moluccus
25½–29½ in | 65–75 cm

This species has been able to take advantage of human settlement, and is common in towns and cities, becoming used to people and hanging around parks and picnic areas looking for a free hand-out or inspecting rubbish bins for scraps. Also common around wetlands, it is a large, distinctive ibis, with a white body and a black head.

Where to find Easily seen in many Top End cities and towns.

Straw-necked Ibis

Threskiornis spinicollis
23½–27½ in | 60–70 cm

This species is common in a variety of open habitats, including wetlands, mudflats, parks, sports grounds, fields and airfields, often around cities and towns. It is a large, easily recognized ibis, dark above and white below with a naked black head and, like the other ibises, a long downcurved bill. Breeding adults have a tuft of bristly, straw-like plumes on their lower neck.

Where to find Large open spaces almost anywhere.

Royal Spoonbill

Platalea regia

29½–31½ in | 75–80 cm

A widespread and easily recognized species, this bird is often seen walking through the shallows, holding its strange black spoon-shaped bill slightly open and sweeping it from side to side, just below the water's surface, as it searches for small fish and invertebrates. A similar species, the **Yellow-billed Spoonbill** *Platalea flavipes* (not illustrated), sometimes occurs in the Top End; it is told by its yellow bill and legs.

Where to find There are nearly always a few birds on Yellow Water in Kakadu NP, but other good places to search include Knuckey Lagoons in Darwin and Fogg Dam.

Glossy Ibis *Plegadis falcinellus*

19½–21½ in | 50–55 cm

This bird is usually found in small to large flocks feeding around shallow wetlands, billabongs and floodplains with exposed mud. It is a small ibis with a long, curved bill and long, dark legs. Adults are dark all over, with a beautiful reddish-green sheen that can sometimes be seen in good light.

Where to find There are usually a few birds on Yellow Water in Kakadu NP, but it can be seen on many of the shallow lagoons across the Top End, including Fogg Dam and Knuckey Lagoons in Darwin.

Where to find The mouth of the Buffalo Creek near Darwin is one of the few readily accessible places where this bird can be seen.

☐ Chestnut Rail *Eulabeornis castaneoventris* 18–20 in | 46–61 cm

This large rail is usually detected by its distinctive loud, braying call. Although quite common, its extreme shyness and the difficulty of accessing its preferred habitat – extensive muddy mangroves – make it a tricky bird to find. To see one, the best strategy is either to sit quietly watching an area of mangroves with a muddy edge, or to cruise quiet mangrove-lined creeks and rivers at low tide looking for birds feeding on the mud. They are lanky, chicken-sized birds with a chestnut-coloured body, grey head and a large yellow bill.

☐ Buff-banded Rail

Gallirallus philippensis

12–13 in | 30–33 cm

This cryptically patterned waterbird is quite common in the Top End, but its secretive habits can make it difficult to see. It is usually seen scurrying around the muddy edges of wetlands, along river banks or on overgrown grassy areas in parks or lush gardens. A largish rail, it can be told from the other species in the region by its finely barred underparts and orangey band across the breast.

Where to find Look for this rail on Yellow Water in Kakadu NP, around Fogg Dam, or anywhere there is rank vegetation near water.

☐ White-browed Crake

Porzana cinerea

7½ in | 19 cm

This small crake is quite common in thick vegetation around lagoons across the northern Top End, but can be difficult to see. The usual sign of its presence is the squeaky call, which sounds like someone repeatedly squeezing a rubber duck! It has a grey breast and head with an obvious white eyebrow.

Where to find Common at Knuckey Lagoons in Darwin and Fogg Dam, and can sometimes also be seen at Yellow Water in Kakadu NP.

Where to find There are usually a few birds present on Yellow Water and Mamukala in Kakadu NP, and at other wetlands including Knuckey Lagoons in Darwin and Fogg Dam.

☐ Purple Swamphen *Porphyrio porphyrio* 17½–19½ in | 45–50 cm

Very common in southeastern Australia, this unmistakable bird is less common in the Top End, but still occurs regularly, particularly in the dry-season. It is usually seen foraging singly or in small parties around the margins of wetlands and lagoons, often close to reedbeds or dense grass. If disturbed, birds generally run quickly for cover, but in some areas get used to disturbance and even become quite tame. This species has also adapted well to human settlement, occasionally being seen feeding on lawns in urban parks where there are ponds or artificial wetlands. It has a varied diet, feeding on invertebrates, small animals and aquatic vegetation. One of its favoured foods is the fleshy bases of grasses and sedges, which it often uproots, holding the stalks in its large spindly feet and feeding delicately. It is a large, dark purple waterbird with a stout red bill and distinctive red shield on the forehead, and has the curious habit of continually flicking its short tail as it forages, revealing a patch of white feathers underneath.

Where to find Most frequently seen at the Leanyer and Palmerston sewage ponds in Darwin.

☐ Common Coot *Fulica atra* 13½–15 in | 34–38 cm

(Eurasian Coot)

This distinctive waterbird often occurs as rafts of thousands of birds in southern Australia, but is less abundant in the Top End, usually only being seen in small groups. However, it is still widespread and may be found on most wetlands, although it tends to favour more permanent water bodies such as man-made dams and sewage ponds, and is less common near the coast. It rarely comes onto land, and is usually seen swimming around on open water, diving for the aquatic vegetation upon which it feeds. Adults are easily recognised by their wholly black plumage and white bill and 'shield' on the forehead, but immature birds have a white face and breast.

Where to find May be seen on almost any open grassy area across the Top End. They are fairly common on the plains surrounding Yellow Water in Kakadu NP, and can sometimes be seen at Knuckey Lagoons or Holmes Jungle Swamp in Darwin.

☐ Brolga *Antigone rubicunda* 67–90 in | 170–230 cm

This iconic Australian bird is more common in the Top End than almost anywhere else in Australia. It is best known for its dancing displays, with birds leaping around flapping their wings and thrusting their heads skyward while giving a loud trumpeting call. The Brolga is usually seen in loose groups or flocks on open plains, around the edges of large wetlands, or sometimes on open ground that has been recently burnt, feeding on invertebrates, roots or aquatic vegetation. After breeding, which occurs during the wet-season, it is often seen in small parties comprising a pair and their offspring. A large, tall crane, it has a pale grey body, red head and long, dark grey legs; immature birds have much duller red heads than adults.

Where to find Not common anywhere, but Buffalo Creek, Lee Point and the Nightcliff foreshore are good areas to search.

☐ Beach Stone-curlew *Esacus magnirostris* 21–23 in | 53–58 cm

(Beach Thick-knee)

Although quite large, this cryptically patterned wader can be difficult to find as it is very shy and prefers undisturbed beaches. It often occurs in pairs, which spend much of the day resting quietly on the ground among sand dunes at the back of the beach, often in the shade of a tree, particularly a mangrove. Although sometimes seen feeding during the day, especially at low tide, birds are most active throughout the night and at dusk and dawn, moving out from their roost to search for marine invertebrates on nearby mudflats, estuaries, beaches and rocky reefs. The Beach Stone-curlew is particularly fond of crabs, stalking them and striking them repeatedly with its powerful bill to break them up before eating them. It is a stocky wader and is unmistakable, having long, yellow legs and a large, yellow-and-black bill; the body is mostly brown, with a dark shoulder patch, but the face has a distinctive black-and-white pattern. In flight it has obvious white patches in the outerwings.

☐ Pied Oystercatcher *Haematopus longirostris* 16½–19½ in | 42–50 cm

Strictly coastal, this distinctive wader is usually found in pairs on open beaches and mudflats, where it spends most of its time along the waterline, searching for shellfish, which are prised open with the powerful chisel-like bill. It is a stocky black-and-white wader with pink legs and a long, reddish-orange bill.

Where to find Good places to see this bird include the beaches at Lee Point and Buffalo Creek in Darwin.

Where to find Although uncommon in the region, it may be seen along rocky coasts such as at East Point and the Nightcliff foreshore.

◀ ☐ Sooty Oystercatcher
Haematopus fuliginosus
15½–20½ in | 39–52 cm

Unlike the Pied Oystercatcher which occurs on sandy beaches and mudflats, this bird prefers rocky areas, although the two species do occasionally occur together. Not common in the Top End, this species is similar in size and shape to the Pied Oystercatcher but is entirely sooty black with a reddish-orange bill and pink legs.

☐ Black-winged Stilt
(Pied Stilt; White-headed Stilt)
Himantopus himantopus
13–14½ in | 33–37 cm

These striking, slim black-and-white birds are common and obvious residents on many wetlands across the Top End. They are usually seen foraging in the shallows at the edge of lagoons, striding through the water on their long, red legs (which in flight trail far behind, looking like a long, red tail). Stilts are very vocal, often giving a strange yapping call that sounds like an angry little dog.

▼

Where to find There are usually a few birds present on Yellow Water and other wetlands in Kakadu NP, while closer to Darwin you can see them at Fogg Dam and Knuckey Lagoons.

■ Masked Lapwing *Vanellus miles* 12–14½ in | 30–37 cm
(Spur-winged Plover)

This long-legged, lanky plover is a familiar species across the Top End, with most parks, sports grounds and grassy fields having a resident pair, and it is also found on most wetlands. Sometimes called 'Spur-winged Plover' due to the small yellow spurs on the bend of the wing, this bird is best known for its loud and aggressive behaviour when breeding, with pairs defending their territory vigorously against any intruders. The nest is a shallow scrape in an open area, and the chicks are precocious, able to run around following their parents almost immediately after hatching.

Where to find Almost any open area or wetland across the Top End.

■ Comb-crested Jacana *Irediparra gallinacea* 8–10½ in | 20–27 cm

('Jesus-bird'; 'Lily-trotter')

Common on lagoons and wetlands with floating vegetation across the Top End, these fantastic birds are best known for their ability to walk across lily pads, giving rise to their alternative names. Jacanas are able to do this thanks to their incredibly large feet and very long toes, which spread their weight over a large area. Pairs also nest on floating vegetation, and the chicks are precocious, being able to run about on the lily pads immediately after hatching. The male is responsible for all the parental care, and if danger threatens he will gather the chicks under his wings, and can even run like this with several pairs of tiny legs dangling below! Difficult to see well from the shore, these birds will usually fly away if threatened, with a distinctive fluttering flight and legs dangling below. However, they seem much more accepting of boats and will often allow quite close approach. A beautiful bird, they are white below with brown wings. Adults have a black breast-band and red 'comb' on the forehead; these features are lacking in younger birds.

Where to find Yellow Water and many of the wetlands in Kakadu NP are good places to see this bird, and it is also common at Knuckey Lagoons in Darwin and Fogg Dam.

Red-kneed Dotterel

☐ **Red-kneed Dotterel** *Erythrogonys cinctus* 6½–7½ in | 17–19 cm

A nomadic species, which is sometimes common but often absent, this small plover is usually found on the muddy edges of freshwater wetlands and lagoons. It has a dark hood, black in adults but brown in younger birds, with a distinctive white throat patch.

Where to find Unpredictable, but often seen in the late dry-season on Yellow Water in Kakadu NP and at Knuckey Lagoons in Darwin.

☐ **Red-capped Plover** *Charadrius ruficapillus* 5½–6¼ in | 14–16 cm

This tiny plover is most common on large, open areas near water, particularly around the coast where it is often found along beaches and on mudflats. Adults are plain brown above and white below, and have a distinctive rufous-red cap.

Where to find Can usually be found on the beach at Buffalo Creek and Lee Point in Darwin.

☐ **Ruddy Turnstone** *Arenaria interpres* 9½ in | 24 cm

This species is strictly coastal and prefers rocky areas, but is often found foraging on sandy beaches too. It breeds in the northern hemisphere before migrating south to spend the summer in Australia. An unusual, dumpy wader, it is mottled brown above, with a white breast and black-and-white patterned head. It has a short, sharp, chisel-like bill that it uses to flick over vegetation, tideline debris or stones.

Where to find Can usually be seen with other waders on the beaches at Buffalo Creek and Lee Point, or at the Nightcliff foreshore.

☐ Black-fronted Dotterel *Elseyornis melanops* 6¼–7 in | 16–18 cm

This striking little plover is common on the muddy edges of freshwater wetlands, lagoons and farm dams, and can very occasionally be found along the coast. It has a distinctive black 'V' on the chest.

Where to find Can usually be seen on the Yellow Water cruise in Kakadu NP, and also at Knuckey Lagoons in Darwin.

Black-fronted Dotterel

Red-capped Plover

Female

Male

Breeding

Ruddy Turnstone

Non-breeding

Pacific Golden Plover
Non-breeding
Grey Plover
Grey Plover
Non-breeding
Non-breeding
Lesser Sandplover
Breeding
Greater Sandplover
Non-breeding

The birds shown here are all coastal species, best seen when they congregate in densely packed roosts at high tide. At low tide they spread out on the coastal mudflats and sandflats to feed. They are all migratory, breeding in the northern hemisphere and spending the Australian summer in the southern hemisphere. Like most shorebirds, these species have a dull non-breeding plumage, which is generally what you will see in Australia, and a brighter breeding plumage, which you may see between February and April just before the birds return north to breed. These birds, and many other shorebirds, are often seen with rings or flags on their legs; scientists use these to identify individual birds and collect information about their incredible migrations.

☐ **Pacific Golden Plover** *Pluvialis fulva* 10 in | 25 cm

Unlike Grey Plover, this species is not restricted to the coast, sometimes being seen on open grasslands or around the muddy margins of freshwater wetlands. It is very similar to Grey Plover but is slightly smaller, less bulky and more yellowish-brown. Birds in breeding plumage are black below, with 'golden' spangles above.

☐ **Grey Plover** *Pluvialis squatarola* 12 in | 30 cm
(Black–bellied Plover)

A medium-sized and long-legged plover restricted to coastal areas, this bird is mainly brownish with whitish mottling and a white belly. Birds in breeding plumage, which you may see from February to April, are spectacular, with black bodies and faces bordered by white, and 'silver'-spangled wings. In flight it has diagnostic black 'armpits'.

☐ **Lesser Sandplover** *Charadrius mongolus* 8 in | 20 cm

This species is very similar to Greater Sandplover and the two can be difficult for the beginner to tell apart. Lesser Sandplover is a little smaller and stockier, with a shorter, stubbier bill, and a steeper forehead, but these features are difficult to assess unless they are close to each other for direct comparison. In breeding plumage it has a black mask and bright rufous breast-band.

☐ **Greater Sandplover** *Charadrius leschenaultii* 9½ in | 24 cm

A small but lanky plover, this species is usually seen foraging alone on open mudflats and sandy beaches, but gathers in large numbers at high tide roosts. It is plain brown above and white below, with a longish, stout black bill which is relatively longer than that of the very similar Lesser Sandplover. Like that species, in breeding plumage it has a black mask and bright rufous breast-band.

Where to find From September to April, all of these species are best found either feeding on the open sandy beaches at Buffalo Creek or Lee Point, or roosting on the rocky foreshore at Nightcliff.

Rather than being found around the coast, these species tend to occur on open grassy areas, such as floodplains, airfields, sports grounds, paddocks and agricultural areas. Oriental Plover, Oriental Pratincole and Little Curlew are all migratory, breeding in the northern hemisphere and spending the southern summer across inland northern Australia. They are usually seen in the Top End as they pass through on their migration early in the wet-season. Australian Pratincole is also a migrant, but within Australia. It spends the dry-season in northern Australia, before heading to southern Australia during the wet-season.

☐ **Oriental Pratincole** *Glareola maldivarum*

9–9½ in | 23–24 cm

This species occurs in the Top End as it passes through on migration to the open plains of inland northern Australia, but is very unpredictable, being seen in thousands some years, and not at all in others. A graceful and long-winged wader, it has a distinctive dark 'necklace' around the throat.

Where to find May occur during the early wet-season on almost any open area across the Top End, such as playing fields or airfields, as they pass through on migration.

Oriental Pratincole

☐ **Oriental Plover** *Charadrius veredus* 9 in | 23 cm

A smallish, slender plover, plain brown above and white below, with a brownish breast and white eyebrow. In breeding plumage the breast is rufous.

Where to find Arrives in October when it is often seen on the Nightcliff foreshore, before moving inland through places like Timber Creek and Pine Creek.

☐ **Little Curlew** *Numenius minutus* 12 in | 30 cm

Usually seen in flocks, this medium-sized, fairly uniformly pale brown wader has a smallish but obviously downcurved bill.

Where to find Often found around Darwin in the early wet-season, and may occur on almost any open grassy area, from suburban playing fields to the floodplains at Holmes Jungle and Knuckey Lagoons.

☐ Australian Pratincole *Stiltia Isabella* 9–9½ in | 23–24 cm

Often seen darting around on the ground searching for insects, this graceful, slender wader has very long black-tipped wings and a short tail.

Where to find Can usually be found during the dry-season on the floodplains of the Alligator Rivers, including Yellow Water in Kakadu NP.

Australian Pratincole

Oriental Plover
Non-breeding

Little Curlew

Sandpipers are an incredible group of birds, and include some of the natural world's most impressive endurance athletes. Between April and September they breed in the northern hemisphere, sometimes even inside the Arctic Circle, before migrating to coasts in the southern hemisphere, including Australia's, from October to March. Incredibly, some of these birds make this trip non-stop! In Australia they are usually found feeding on mudflats or sandflats around the coast, or at shallow freshwater wetlands. To the inexperienced observer, many look the same, but with practice you can recognize some of the more distinctive species. This page covers some of the largest and most obvious species you are likely to see.

☐ Whimbrel

Numenius phaeopus

17 in | 43 cm

Usually found singly on open mudflats or rocky reefs around the coast, this medium-sized wader has a longish, downcurved bill. It is mottled brown but has a distinctive white eyebrow and dark stripes on the crown.

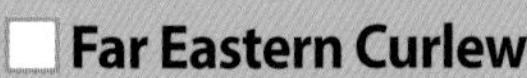

☐ Far Eastern Curlew

Numenius madagascariensis

25 in | 64 cm

Fairly common on open mudflats and beaches, this unmistakable, large mottled brown wader, has an extremely long, downcurved bill.

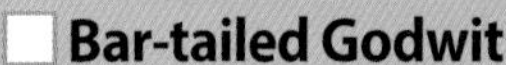

☐ Bar-tailed Godwit

Limosa lapponica

16 in | 41 cm

Quite common on mudflats around the coast, where it is often seen in small flocks probing the mud, this medium-sized wader has a long, slightly upturned bill that is pink at the base and dark at the tip. Late in the wet-season this species is often seen moulting into breeding plumage, with the head and underparts turning deep reddish in colour.

Where to find The beaches and mudflats at Buffalo Creek and Lee Point, as well as the Nightcliff foreshore in Darwin are good places to look for all these species.

Whimbrel
Far Eastern Curlew
Bar-tailed Godwit

Like most of the waders, these species are migratory, present in large numbers over the wet-season. Although most of them return to the northern hemisphere to breed from April to September, a few remain behind each year, including some younger birds, or birds which are not in good enough condition to make the arduous journey to the breeding grounds. Unfortunately, the numbers of waders visiting Australia each year has been steadily or, in some cases such as the Red Knot, drastically declining. The primary reason is the development or disturbance of much of their coastal habitat, not just in Australia but also along their migration routes throughout east Asia.

☐ **Common Greenshank** *Tringa nebularia* 13 in | 33 cm

Usually solitary, this species is fairly common on mudflats and also along brackish creeks or among mangroves, and sometimes on freshwater wetlands. It is an active feeder, often darting about in shallow water chasing prey. A lanky wader, at first glance it appears pale all over, but on closer inspection is white below with pale grey wings. It has longish green legs and a long, slightly upturned bill.

☐ **Red Knot** *Calidris canutus* 10 in | 25 cm

Similar to Great Knot, but less common, this species is usually only seen in small parties in the Top End. It is slightly smaller and dumpier than Great Knot and has less distinct spotting along the flanks, and a shorter, stubbier bill. Like several waders, the name is misleading for Australian naturalists; it is only 'red' in breeding plumage, when the body becomes a deep orange-brown colour. Birds are only likely to be seen in this plumage around March, before they depart on the northward migration to their breeding grounds in the northern hemisphere.

☐ **Great Knot** *Calidris tenuirostris* 10½ in | 27 cm

This species is one of the more common waders around Darwin, and often one of the dominant species in wader flocks. It has a slightly downcurved bill and, in non-breeding plumage, is brown above with black spots or 'arrows' along the flanks. In breeding plumage the breast and neck become heavily spotted with black; birds can often be seen in this plumage in the late wet-season, just before they migrate north to their breeding grounds in northeast Siberia.

☐ **Grey-tailed Tattler** *Tringa brevipes* 10½ in | 27 cm

Perhaps the plainest of the waders, this species tends to be quite solitary, usually foraging alone, although at high tide roosts it often gathers in small flocks. It is grey all over apart from an indistinct white eyebrow, and has yellowish legs and a stout, straight, blackish bill. Even in flight, when most species show white wingbars or rumps, the tattler is just plain grey.

Where to find Any of these species can be found on the beaches and mudflats at Buffalo Creek and Lee Point in Darwin.

Greenshank

Non-breeding

Red Knot

Breeding

Great Knot

Non-breeding

Breeding

Grey-tailed Tattler

Sometimes called 'peeps', due to the sounds they make while feeding, these little waders are a birdwatcher's nightmare, being small, difficult to see well, and all looking very similar! They are also some of the most incredible athletes on the planet – Red-necked Stints weigh little more than a matchbox, yet migrate more than 10,000 km twice a year. They are usually seen on coastal mudflats and beaches, scurrying around at the edge of the water as the tide recedes and feeding on tiny marine invertebrates.

Where to find Any of these species can be seen on the coastal mudflats and beaches at Buffalo Creek and Lee Point in Darwin.

☐ Sanderling *Calidris alba* 8½ in | 22 cm

Usually seen in small flocks at the beach, running backwards and forwards in unison with the waves, feeding on small invertebrates on the exposed sand. Another supreme athlete, this species migrates north to the Artic Circle and back each year to breed. It is similar to the Red-necked Stint, but generally whiter below and paler, silvery-grey above, and has a slightly longer bill.

☐ Red-necked Stint
Calidris ruficollis 6 in | 15 cm

This tiny, stocky little wader has a small, fine black bill. The name is confusing, as they only have a red neck in breeding plumage; for most of their time in Australia they are pale grey above and white below, with a dusky grey collar.

Sanderling
Non-breeding

Red-necked Stint
Non-breeding

Curlew Sandpiper

Calidris ferruginea

8 in | 20 cm

Found mostly around the coast, this shorebird often wades deeper into the water when feeding than other similar species. Small but lanky with a longish, black, downcurved bill, like most of the waders seen in Australia during the southern hemisphere summer it is plainly coloured, but birds start to moult into their deep reddish breeding plumage just before they return north in around March.

Terek Sandpiper

Xenus cinereus

9½ in | 24 cm

These distinctive, dumpy little shorebirds with a long, upturned bill and short, yellow legs dart about on the mudflats like dinky little wind-up toys.

Other small sandpipers

While most shorebirds are restricted to, or at least more common on the coast, the species on this page can also be found on freshwater wetlands or more sheltered sites. Each of these species is migratory, and while small numbers may be present year-round, most only occur in the Top End between September and April.

☐ **Marsh Sandpiper** *Tringa stagnatilis* 9½ in | 24 cm

This species breeds across central Russia and northern China, south of the Arctic Circle, with the majority of the population migrating to Africa for the southern hemisphere summer, although some birds do migrate to Australia. Arriving around October, it is fairly common across the Top End, generally being found in ones and twos on the margins of shallow freshwater wetlands and lagoons. An active feeder, it moves around quickly and will often wade into deep water when searching for food. A small, slender shorebird, it appears pale, with white underparts and greyish-brown wings and crown, looking like a small, delicate Common Greenshank (*page 66*), but has a relatively straight and thinner needle-like bill.

☐ **Common Sandpiper** *Actitis hypoleucos* 8 in | 20 cm

A solitary bird, this species is usually seen scurrying around on the mud, continually stopping and teetering up and down before running off again. It prefers sheltered mudflats or rocky areas and is usually seen along the edges of smaller creeks and rivers, quite often around mangroves. Unlike other waders, which are usually only seen on mudflats, the Common Sandpiper is often seen perched on logs, fallen trees or posts at the water's edge. It is a small, rather dumpy little wader, brown above and white below, with a short, straight bill, and when flushed flies with distinctive stiff, downcurved wings that are flicked in short bursts.

☐ **Sharp-tailed Sandpiper** *Calidris acuminata* 7½ in | 19 cm

Quite common in the Top End, this sandpiper often occurs in loose flocks that forage on the shallow muddy margins of freshwater wetlands, but is equally at home on coastal mudflats and sometimes beaches. From April to September it breeds inside the Arctic Circle in far northeastern Siberia, but nearly the entire population migrates to Australia for the southern hemisphere summer. A smallish shorebird with yellowish legs and a slightly drooping bill, this species has a distinctive rufous cap that can be seen with good views. It is less uniform above than many other waders, with each of the feathers on the back having dark brown centres and paler edges, giving it a mottled or blotchy appearance.

Where to find Common Sandpiper is often found along canals and mangrove-lined creeks at sites like Buffalo Creek in Darwin, while Marsh Sandpiper can be seen on freshwater wetlands like Yellow Water in Kakadu NP and Knuckey Lagoons in Darwin. Sharp-tailed Sandpiper is possible at any of these sites.

Marsh Sandpiper
Common Sandpiper
Sharp-tailed Sandpiper

☐ **Silver Gull** *Chroicocephalus novaehollandiae* 17–17½ in | 43–44 cm

The common 'seagull' throughout Australia, Silver Gulls are best known for harassing picnickers at the beach for chips or other scraps, with their antics giving rise to the common Australian saying "like seagulls fighting over a chip!" Watching their interactions in these situations can be fascinating. There is usually one dominant bird that assumes a number of characteristic postures, including craning its neck or lowering its head close to the ground while calling loudly, threatening other birds that come too close. However, getting too absorbed in this behaviour often means it misses the chase for scraps! These birds are most abundant near the coast, but have adapted well to human settlement and are common around rubbish tips, or anywhere they might find food. Mostly white with silvery-grey wings, the bill and feet are bright red. Younger birds may have some brown mottling on the wings and a brown bill.

Where to find Anywhere along the coast and very easy to see in Darwin, particularly at Stokes Hill Wharf.

☐ **Caspian Tern** *Hydroprogne caspia* 18½–21 in | 47–53 cm

This large tern is found around coasts, rivers and lakes across the world, and is fairly common in Australia. It is most regularly seen around beaches and estuaries, but unlike the other large terns is equally at home on freshwater wetlands or along rivers. It spends its time resting on the sand in ones and twos, often in mixed groups with other terns or gulls, or patrolling over the water searching for fish, which it catches by plunging headfirst. It is much larger than other tern species, and the huge red bill makes it easy to identify. In breeding plumage the cap is solid black, but in non-breeding birds it is streaked with white.

Where to find Around Darwin there are usually a few at Buffalo Creek, Lee Point, and sometimes also at Knuckey Lagoons.

Non-breeding

Non-breeding

Lesser Crested Tern
Thalasseus bengalensis
15–17 in | 38–43 cm

This tern is found around the coast, mainly during the late dry-season, often resting in small flocks, or mixed in with other terns. It is very similar to the Greater Crested Tern but slightly smaller and more slender, with a finer, more orange-coloured bill.

Where to find Sometimes found at Buffalo Creek and Lee Point in Darwin.

Greater Crested Tern *Thalasseus bergii*
15½–19½ in | 39–50 cm

This tern is quite common around the coast, along beaches, reefs and around estuaries. They usually rest in small groups, often with other tern species mixed in. They are acrobatic feeders, plunging into the water in steep dives to catch small fish. They have a slender, yellowish bill and a solid black cap and crest, which becomes streaky when not breeding.

Lesser Crested Tern
Breeding

Non-breeding

Greater Crested Tern

Non-breeding

Breeding

Where to find Good places for this bird include Lee Point, Nightcliff and Stokes Hill Wharf in Darwin.

Where to find Stokes Hill Wharf in Darwin is a good place to see this tern, and also to compare it with some of the other terns that occur in the area. It can also be seen at Buffalo Creek and Knuckey Lagoons in Darwin.

☐ Gull-billed Tern *Gelochelidon nilotica* 15–17 in | 38–43 cm

This medium-sized tern is quite common across the Top End, with a few individuals found on most large bodies of water. There are two subspecies that occur in the Top End, the resident Australian birds (which some scientists treat as a separate species, the **Australian Gull-billed Tern** *Gelochelidon macrotarsa*), and a smaller subspecies which is a migrant from Asia. It is best recognized by its stout, black bill. It has a solid black cap in breeding plumage, but outside the breeding season this is moulted out, leaving only a small dusky ear patch.

Non-breeding

Non-breeding

White-winged Tern
Breeding
Non-breeding
Whiskered Tern
Non-breeding
White-winged Tern
Non-breeding
Little Tern
Whiskered Tern
Non-breeding
Breeding

☐ White-winged Tern

Chlidonias leucopterus
9–9½ in | 23–24 cm

This migratory species is only present during the wet-season, and is often found in the same places as Whiskered Tern, but in smaller numbers. A small tern that can be difficult to identify, it is long-winged and short-tailed, with a short black bill and dusky 'helmet', with dark patches on the cap and behind the eye. As this species comes into breeding plumage (around February) the body gradually becomes completely black and the shoulders white.

Where to find The sewage ponds in Leanyer and Palmerston are good places for this species. There are also usually a few birds around Stokes Hill Wharf.

☐ Little Tern

Sternula albifrons
8–10 in | 20–25 cm

This tiny tern is usually seen flying along the beach with its distinctive bouncy flight, often twisting and plunging vertically into the water in pursuit of fish. In non-breeding plumage it has a dark bill and dusky nape, but when breeding the bill turns yellow and the cap becomes solid black, contrasting with a white forehead.

Where to find Around Darwin there are often a few birds mixed in with the tern flocks at Buffalo Creek and Lee Point, and it is also regular at Stokes Hill Wharf.

☐ Whiskered Tern

Chlidonias hybrida
9½–10 in | 24–25 cm

Unlike other terns, this species is most common on freshwater wetlands across the Top End. It often occurs in small flocks, and is usually seen flapping around gracefully over the water, sometimes stooping down to pick food off the surface. Easily identified in breeding plumage, when it has grey wings, a grey belly, black cap, white cheeks and a dark red bill, when in non-breeding plumage its belly is white, the bill is black, and the cap becomes streaked.

Where to find Easily seen on Yellow Water in Kakadu NP; regular at Knuckey Lagoons in Darwin and Fogg Dam.

☐ **Azure Kingfisher** *Ceyx azureus* 6¼–7½ in | 16–19 cm

This stunning little kingfisher is common on waterways across the Top End, but can be difficult to see well. It is usually seen streaking past in direct flight, low over the water like a little blue missile. Its passing is often announced by a high-pitched "*seep*" call which, if learnt, will give you a good clue to the bird's presence. A good strategy for seeing this species is to sit quietly beside a small creek or billabong where they are known to occur. With patience, and a little bit of luck, you may see a bird perched quietly at the water's edge, usually on a low branch or snag, staring intently at the water and constantly bobbing its head (which helps with depth perception). If it spots a fish, it will dive off its perch and plunge into the water in pursuit, usually returning to the same perch with a tiny fish in its bill. This beautiful bird is dark purplish-blue above and orange below, with tiny red feet and a long, black bill.

Where to find The Yellow Water cruise in Kakadu NP is a great place to see this little gem, although it can be found on most waterways.

☐ **Little Kingfisher** *Ceyx pusillus* 4½–5 in | 11–13 cm

These lovely little kingfishers are a Top End specialty, being seen more easily here than elsewhere in northern Australia. Like Azure Kingfisher they can be tricky to see well as they are generally quite shy; if not seen zipping along low over the water, they may be found perched on a sheltered snag, often in the shade of a low bush or overhanging tree. Although they generally fly away at the first sign of disturbance, individuals can be quite approachable by boat. During the wet-season these birds become very difficult to find as pairs begin to breed. They are so secretive at this time that very little is actually known about their nesting habits, other than that they nest in a small tunnel which they excavate in the bank of a creek or among the roots of a tree, and that the young become independent very soon after leaving the nest. One of the smallest kingfishers in the world, Little Kingfishers have a blue head and wings but, unlike the similar Azure Kingfisher, they are white below and have black feet.

Where to find Most often seen on Yellow Water in Kakadu NP, it can also be seen very occasionally from the boat ramp at Buffalo Creek in Darwin.

Raptors
Hawks, Eagles and Falcons

White-bellied Sea-eagle

Black Kites feeding over a fire

There is something about raptors and their mastery of the air that captures the imagination and they are a group of birds that everyone enjoys watching. Birds of prey are found across the landscape in all environments and have evolved to fill many niches with a corresponding variety in their natural histories. Each species is specialized to some degree, depending on its preferred food and particular foraging technique. Some birds have adapted to spending much of their time soaring on broad wings searching for carrion (dead animals), while others are small and manoeuvrable, specializing in the high-speed pursuit of other birds.

Wedge-tailed Eagle

Where to find Regularly seen near Darwin: Lee Point and the Nightcliff foreshore are good places to look.

☐ **Osprey** *Pandion haliaetus*
19½–25½ in | 50–65 cm

This raptor feeds exclusively on fish, so is always found near water. It is fairly common around the coast, over beaches, reefs and estuaries, but sometimes moves up large rivers and may occasionally occur far inland. Ospreys build large, bulky nests that are often exposed in the tops of dead trees or on power pylons, and may be used repeatedly for years. In flight, the pale underside, long wings and relatively short tail are good identification features.

Where to find Often seen over suburban Darwin, but most regular around estuaries like Buffalo Creek or the Nightcliff foreshore.

☐ Brahminy Kite *Haliastur indus*

17½–19½ in | 44–50 cm

Quite common around the coast, and sometimes found a little way up larger rivers, this beautiful raptor is regularly seen soaring along the shoreline. It feeds mostly on fish and carrion, and is particularly fond of dead fish that have been washed up on the shore. Adults have rich reddish-brown wings with a striking white breast and head, while younger birds are brown all over and streaky, making them difficult to separate from Whistling and Black Kites (*pages 88–89*).

Immature

Adult

Where to find There are resident pairs at Yellow Water in Kakadu NP, and this is a great place to see them. Other good spots include the coast around Darwin and, in the southwest, Victoria River near Timber Creek.

☐ White-bellied Sea-eagle *Haliaeetus leucogaster*

29½–33½ in | 75–85 cm

This majestic eagle is usually seen soaring over the coastline or large wetlands, where the distinctive steep 'V'-shape of the wings and short tail are obvious. It feeds on almost anything it can catch, including fish, turtles, snakes, lizards and birds, as well as carrion. When nesting, pairs usually repair an old nest, a huge, bulky pile of sticks, often in the top of a large tree, which may be used many years in a row. Adult birds are white below and blue-grey above with a black trailing edge to the underwing, while younger birds are tawny-brown all over, gaining adult plumage over about four years.

☐ **Pacific Baza** *Aviceda subcristata* 15–17½ in | 38–44 cm

Uncommon in the Top End, this beautiful raptor may be found in savanna woodlands and also in gallery forest along waterways, usually in pairs or small family groups. It preys mainly on small vertebrates such as frogs and lizards, but also has a particular fondness for stick insects. Bazas catch their prey by plucking it from the canopy of trees, and sometimes try to flush prey by flapping around awkwardly in the outer foliage. Usually seen soaring overhead, this species' distinctive underwing pattern and broad paddle-shaped wings make it fairly easy to identify in flight. When seen perched it is unmistakable, having a blue-grey head, dark grey wings and white breast with brown barring. It also has large, yellow eyes, and unlike any other Australian raptor a small crest at the back of the head. Young birds are similar to adults but are browner.

Where to find Seen regularly around suburban Darwin, Kakadu NP, Katherine and Mataranka.

☐ **Wedge-tailed Eagle** *Aquila audax*

33½–42 in | 85–107 cm

Australia's largest raptor and one of the largest in the world, this huge bird is quite common in the Top End, where it is usually seen soaring in pairs, or standing by the road feeding on carrion. It is rarely seen perched as it is quite shy, and will fly away at the first sign of danger. Sometimes if one is flushed up from the side of the road it will perch in a roadside tree which provides a good opportunity to see it well. Wedge-tailed Eagles can be confused with young White-bellied Sea-eagles (*page 84*) which are also brownish – always look for the tail, which is long and wedge-shaped in this species, but shorter in the sea-eagle. The sea-eagle also holds its wings in a much deeper 'V' when soaring. 'Wedgies', as they are sometimes referred to, maintain huge nests which may be used repeatedly for years. Interestingly, the female always lays two eggs, usually a few days apart, but often only the first chick to hatch survives as it is able to outcompete the younger chick for food. The adults may occasionally raise both chicks, but only in good seasons when there is plenty of food available. Young eagles are reddish-brown, with paler patches on the back of the neck and shoulders. As they get older they become darker, and very old adults may be almost black.

Where to find Most often seen in the drier parts of the Top End, around Katherine and along the highway southwest to Victoria River Crossing and Timber Creek.

Where to find Occasionally seen in Kakadu NP, but most common around Katherine and the drier areas to the southwest.

☐ Black-breasted Buzzard

Hamirostra melanosternon

21½–23½ in | 55–60 cm

This distinctive raptor can be seen in most habitats across the Top End, but particularly favours the savanna woodlands and rocky areas of the south and southwest. As with many birds, there is a general movement northwards in the dry-season, and at that time of year this species is occasionally seen close to Darwin. Resembling a small Wedge-tailed Eagle, this large raptor is usually seen in flight, where the obvious features are the short tail and long wings with a white patch in the primaries (outer wing feathers). It also holds its wings in a steeper 'V' than the eagle, and often looks unsteady in flight, rocking from side to side as it glides along.

☐ **Black Kite** *Milvus migrans* 17½–21½ in | 44–55 cm

One of the most frequently asked questions about birds from visitors to the Top End, is "What are those big, brown hawks I see along the highway?" They can be either Whistling or Black Kites, and both are often seen together, although Black Kite is much more common. They often gather in groups, and it is not uncommon to flush ten or more birds fighting over a carcass by the roadside. Black Kites are particularly numerous in urban areas, often being found around rubbish dumps, farms and cattle yards where there is ample food for them to scavenge. At the Darwin rubbish dump they can sometimes be found in their thousands. Black Kites are regular attendants around small grass fires, where they will prey on small animals and insects fleeing the fire front. There are even reports of birds picking up glowing embers and dropping them ahead of the fire to ensure it keeps burning! Black Kites are quite variable in colour, from almost chocolate to much paler brown. The best identification feature is the forked tail; Whistling Kites have a rounded tail.

☐ **Whistling Kite** *Haliastur sphenurus*

19½–23½ in | 50–60 cm

Less common than the very similar Black Kite, both species are often seen gathering around roadside carcasses, although Whistling Kites are usually outnumbered by Black Kites. Whistling Kite occurs right across the Top End but tends to be seen more regularly near water than Black Kite, often around wetlands and along rivers – but wherever one species occurs, the other is also possible. The two species can be difficult to tell apart, but the most obvious difference is the tail: Whistling Kites have a rounded tail, whereas Black Kites have a forked tail. Whistling Kites are also generally paler and more mottled looking, and have a distinctive underwing pattern with a pale patch towards the end of the wing.

Adult

Immature

Where to find The coastline around Darwin, Yellow Water in Kakadu NP or any of the large rivers are good places to look.

☐ Brown Goshawk ▼

Accipiter fasciatus
17½–21½ in | 44–55 cm

Although quite common in the Top End, this secretive raptor can be difficult to see except when it is soaring, often in the morning. It hunts other birds, so waterholes where honeyeaters, finches and other small birds gather in the morning and afternoon are good places to look for one. It is a sleek, brown raptor with finely barred underparts and a long, rounded tail. The very similar but less common **Collared Sparrowhawk** *Accipiter cirrhocephalus* (not illustrated) is almost identical, but is slightly smaller and has a long, square-cornered tail.

Where to find Possible in all habitats across the Top End.

Where to find There are a few pairs around Darwin, and it may also be seen along the rivers in Kakadu NP

☐ Grey Goshawk ▶

Accipiter novaehollandiae
16–22 in | 41–56 cm

Like the other goshawks, Grey Goshawk is a secretive species, most easily seen mid-morning when it often soars above the treetops. It favours wet habitats, such as monsoon and lush gallery forests, particularly along large rivers. There are two colour-morphs: one with pale grey upperparts and finely barred underparts; and another that is pure white.

Where to find Sometimes seen in Kakadu NP, the Daly River Region and also around Elsey NP near Mataranka.

☐ Red Goshawk *Erythrotriorchus radiatus*

19–24 in | 48–61 cm

Ask any birdwatcher visiting the Top End to rank their most wanted species and this one will probably be at the top of their list! Unless you are specifically searching for it you are unlikely to see one as it is among Australia's rarest birds, but the Top End is something of a hotspot for the species. It seems to occur most regularly in extensive tall savanna woodlands near large rivers, and like other goshawks, it hunts mostly birds, sometimes catching prey as large as kookaburras. This bird is very shy, sitting quietly in the foliage of trees or flying stealthily through the canopy in search of prey, and hence it is very difficult to see.

Where to find Can be seen in dry, open habitats and farmland.

Brown Falcon *Falco berigora*

15½–19½ in | 39–50 cm

This large, lanky falcon is commonly seen by the roadside, sitting in dead trees or on fence posts. It has a varied diet, and will eat almost anything it can catch, including insects, lizards, snakes, small mammals and birds. This species will often attend fires, where it mixes with Black and Whistling Kites (*pages 88–89*) – look for the broader, more pointed and upswept wings of the falcon to separate it from the more abundant kites. Quite variable in plumage, in the Top End most birds are brown above and pale below, but some, particularly younger birds, are dark all over.

Brown Falcon

Where to find Look for this species around the escarpments of Kakadu NP, where there is often a pair at Nourlangie Rock.

Peregrine Falcon

Falco peregrinus

15–19½ in | 38–50 cm

Perhaps the world's best-known falcon, this species is found on every continent except Antarctica. Although it occurs in all habitats, it is most often found around cliffs and escarpments where it nests on rocky ledges. It is a powerful, stocky falcon, blue-grey above and pale below with a black hood, and has pointed wings and a distinctive flickering flight.

Peregrine

Australian Hobby

Falco longipennis 12–14 in | 30–36 cm

A beautiful little falcon that is often seen perched on telephone poles, dead trees or tall buildings, or hawking for insects late in the afternoon around small waterholes, parks and gardens. It is small and slim with long, sharply pointed wings, and is much more slender than the similar Peregrine Falcon.

Where to find Often seen around urban areas such as Darwin and Katherine.

Nankeen Kestrel

(Australian Kestrel)
Falco cenchroides
12–14 in | 30–36 cm

This graceful, slender falcon is quite common across the Top End, and is usually seen perched on telephone wires or fence posts along roadsides. Like kestrels all over the world, it often hovers, hanging suspended in mid-air while searching the ground for insects and lizards.

Where to find May be found in any open habitat including farmland.

Birds of the Forest
Mangroves and Monsoon Forests

Rose-crowned Fruit-dove

The term forest is loosely defined, but is generally used to describe more densely wooded habitats. Forests are often associated with wetter microclimates, and this is the case in the Top End where there are two primary types: mangrove forests and monsoon/gallery forests. Mangrove forests are widespread around the coast, estuaries and large tidal rivers, and hold a suite of birds that are found only in this habitat. Monsoon and gallery forest occurs in patches across the Top End: in sheltered areas along the coast, occasionally at the base of sandstone escarpments, and often along the edges of rivers and smaller creeks. Many of the species included in this section are found in both mangroves and monsoon forest, whereas some are specialists of one or the other.

Rainbow Pitta

☐ **Orange-footed Scrubfowl** *Megapodius reinwardt* 16½–18½ in | 42–47 cm

The loud crowing and cackling calls of these birds are often given at dusk, and are a common sound at night across suburban Darwin. Preferring wetter habitats, they are found in monsoon forests, lush parks and gardens and sometimes mangroves, usually near the coast. Nearly always found in pairs, they spend much of their day scratching around in leaf-litter searching for fruit and roots, but spend the night roosting high in a tree, often near water. The scourge of many a Darwin gardener, these birds are like small bulldozers, scratching away at everything in their path, regardless of whether it is the forest floor or someone's well-manicured garden! They are best known for the huge mounds of leaf-litter – these may be many metres in diameter – that each pair forms by scraping with their powerful feet. The mound essentially becomes a large compost heap, with the heat generated by the rotting vegetation being used to incubate the eggs.

Where to find Good places to look around Darwin include the Botanic Gardens, Buffalo Creek and Howard Springs, where a pair often maintains a nest near the car park.

Both birds carefully monitor the temperature of the mound's interior by digging a number of holes and testing the conditions. When the female is satisfied, she lays an egg in one of these holes, covers it up and leaves it to incubate; she will lay around six eggs in the mound each breeding season. The young birds are the size of baby chickens when they hatch, and are much more developed than most other bird hatchlings, being able to dig their way out of the mound before disappearing into the forest. They are able to fend for themselves immediately and do not receive any care from their parents. Unlikely to be confused with any other species, as the name suggests, Orange-footed Scrubfowl have powerful orange legs and feet. They are large, plump birds, the size of a large chicken, which have a grey body, brown wings and a small crest.

Where to find Good places to search include Howard Springs and East Point near Darwin, and around the galleries at Nourlangie Rock in Kakadu NP.

☐ Brown-capped Emerald Dove *Chalcophaps longirostris* 9–11 in | 23–28 cm

(Pacific Emerald Dove)

A ground-dwelling dove, this species spends much of its time wandering around on the forest floor searching for food. It is generally shy and rarely ventures far from cover, although can be quite confiding if you keep still and are patient. It is most often seen trundling around the edges of clearings or trails in the forest, or along the edges of roads at dawn and dusk. Often the first sign one is present is when it flushes suddenly from the side of the trail, seemingly from beneath your feet, and giving quite a fright! If you can follow its rapid flight through the trees it will sometimes perch to look back at you before returning to the ground or flying off. Quite a small, dumpy pigeon, this bird has a pinkish-brown body and emerald-green wings, with two pale grey bars across the rump.

Where to find Parks and gardens in Darwin, and also in monsoon forest and other forested habitats south to Katherine.

☐ Torresian Imperial-pigeon *Ducula spilorrhoa* 15–16½ in | 38–42 cm

This large and distinctive pigeon is a common sight in suburban areas from Darwin to Katherine, and is often seen flying overhead in small flocks, feeding in fruiting trees or sitting on power lines. Some birds are present all year, but the majority of the population are migrants, arriving in Australia from New Guinea to breed around August–September and staying until April. They build scanty little nests, and several will often nest in the same tree. It is a striking bird, being mainly white with some black in the wings and a black tail.

☐ **Black-banded Fruit-dove** *Ptilinopus alligator* 13–16½ in | 33–42 cm

(Banded Fruit-dove; Grey-rumped Fruit-dove)

This beautiful pigeon is one of only a handful of birds found in the Top End but nowhere else in the world, making it a top target for many of the international birders who visit each year. It has very specific habitat preferences, being restricted to monsoon forests at the base of Arnhem Land's sandstone escarpments, and there are only a handful of accessible sites where it can be seen. It is an unobtrusive species, spending most of the day feeding quietly or resting in fig trees, and is difficult to find except in the early morning or late afternoon when birds can be seen flying around at the base of the escarpment, to and from their feeding trees. It has a white head and dark grey back and is easily identified when you see the black band between the white breast and grey belly.

Where to find Nourlangie Rock in Kakadu NP is the most reliable site. Walk quietly among the galleries and look for them feeding in fruiting trees, or sit at the Gunwarddehwarde Lookout in the morning or afternoon and watch for them flying back-and-forth along the escarpment.

☐ Rose-crowned Fruit-dove *Ptilinopus regina* 8½–10 in | 22–25 cm

Although quite common in monsoon forest across the northern Top End, this stunning, small pigeon is unfortunately very difficult to see. The best clue to its presence is the distinctive call, a deep and accelerating "*hoo........hoo......hoo.... hoo...hoo..hoo-hoo-hoo*", although having tracked down the call, you then have to find a small green bird sitting still among a mass of green leaves – no easy feat!

If you are able to find a fruiting fig tree in the forest, you may be lucky enough to see several pigeons arrive and clamber around among the foliage, gorging on fruit. Males of this beautiful species have a pale grey head, green wings, pink breast and orange belly, and a rose-pink crown bordered by a thin yellow line. Females are similarly patterned but are slightly duller.

Where to find Good places to search for this bird are East Point and Howard Springs near Darwin, and the monsoon forests at Fogg Dam.

☐ **Brush Cuckoo** *Cacomantis variolosus* 9 in | 23 cm

When the demented, continuous calling of this cuckoo begins, it is a sure sign the wet-season is on its way. The call is a distinctive descending and accelerating "*pee-ew, pee-ew, pee-ew, pee-ew, pee-ew*", and continues *ad nauseam*, day or night. Like most cuckoos, Brush Cuckoos are brood parasites, laying their eggs in other birds' nests, usually those of smaller birds like honeyeaters and flycatchers. When it hatches, the cuckoo chick ejects the other eggs or chicks from the nest, leaving the unsuspecting 'parents' to raise a single voracious youngster, which will eventually grow to be twice their size! Common and easy to hear, they will often come to investigate a simple whistled imitation of their call.

Where to find May be found in wooded habitats, parks and gardens almost anywhere.

☐ Little Bronze-cuckoo *Chrysococcyx minutillus* 6 in | 15 cm

Fairly common in wooded habitats across the Top End, particularly monsoon forests, mangroves, parks and gardens, this is another bird whose call is the best clue to its presence – a distinctive, descending 5–6-note whistle that is easily imitated. This small cuckoo parasitizes small birds, mainly Large-billed and Mangrove Geryones (*pages 110–111*) in the Top End. The gerygones and cuckoos are in a constant 'arms race', as the gerygones evolve traits to prevent parasitization, while the cuckoos evolve ways to defeat these traits. Fascinatingly, gerygone chicks have evolved a distinctive pattern of white feathers on their otherwise naked backs to help their parents detect and remove cuckoo chicks; and incredibly the cuckoo chicks have evolved to imitate this pattern. Although common, this small, emerald-green cuckoo can be difficult to see well, as it spends much of its time high in the canopy.

Where to find Good places to look include Buffalo Creek, Lee Point, East Point and Howard Springs near Darwin, and Fogg Dam.

Where to find Howard Springs near Darwin is a good place to look for this species, and Bitter Springs near Mataranka is another regular haunt.

Rufous Owl *Ninox rufa* 18–22 in | 46–56 cm

These huge and spectacular owls are widespread, but nowhere common across the Top End. They are usually found in monsoon forest, particularly near watercourses, and sometimes in suburban parks and gardens. During the day, the birds roost in thick vegetation, making them difficult to find, although they will return regularly to the same roost for a period. There are sometimes known roost sites, particularly around Darwin, and this offers the best chance of seeing one. Rufous Owls feed on large mammals, with flying-foxes a primary food source, and they are sometimes found roosting with a half-eaten flying-fox clasped in their powerful talons.

Where to find Quite common around Darwin and can often be seen at East Point, Howard Springs and Buffalo Creek.

■ Large-tailed Nightjar *Caprimulgus macrurus* 11 in | 28 cm

These small nightjars are found right across south-east Asia, and are fairly common in the northern Top End, usually close to the coast. Occurring mainly in monsoon forest and sometimes lush parks and gardens, they are best known for their unusual call, a repetitive "*chonk, chonk, chonk, chonk, …*", often likened to the sound of someone chopping wood. Like other nightjars they spend the day sitting quietly in leaf-litter, where their cryptically patterned plumage makes them very difficult to see – and they generally go unnoticed until they flush, usually just before you are about to step on one! They emerge at dusk to hawk for insects around the edges of the forest, or over open areas and water, and this is the best time to see them. Search for them with a spotlight around the edges of any clearings, or if you hear one calling, looking out for the bright red eye-shine. They have a very buoyant, fluttering flight and have obvious white corners to the tail – the key to separating them from the larger Spotted Nightjar (*page 141*) which is usually found in much drier habitats.

Where to find Good places to search for this bird include East Point and Howard Springs in Darwin, Fogg Dam and the rock art galleries at Nourlangie Rock in Kakadu NP.

☐ Rainbow Pitta *Pitta iris* 6½–7 in | 16–18 cm

Pittas are a famously beautiful and famously shy family of birds with a penchant for living in dense, dark rainforests throughout Asia and northern Australia. This makes them a major target for visiting naturalists, but also one of the most frustrating families of bird to see well. Fortunately, the Rainbow Pitta is probably the easiest of them all to see, and indeed they can be found without much difficulty within minutes of the Darwin Central Business District. The only species of pitta that is restricted to Australia, these birds are common in monsoon forest in the northwestern Top End, being found from Darwin south to Kakadu NP. They spend most of their time hopping around on the ground and scratching in the leaf-litter with their powerful feet in search of their two preferred food items: worms and snails. They call often, and during the breeding season from September to February can be heard day and night – an easily imitated "*we-wit, we-wit*", usually given from a slightly elevated perch.
The males and females of this beautiful bird are alike, having a jet-black body, emerald-green wings and tail, sky-blue shoulder patches and a patch of red under the tail.

☐ Collared Kingfisher *Todiramphus chloris* 9–10½ in | 23–27 cm

This large kingfisher occurs exclusively along the coast, where it is quite common and nearly always found in or close to mangroves. It feeds on fish, crabs and other marine invertebrates, which it catches by diving into the water, or by picking them off mudflats and rocky reefs. Quite a vocal species, it is often detected by its strident call, a three- or four-note "*kek-kik-kik*", with the first note often a little softer than those that follow. Often confused with the very similar but smaller Sacred Kingfisher (*page 143*), Collared Kingfisher has a white breast and dark green wings, while the Sacred has a buffy breast and turquoise wings.

Where to find Good places to look include the Nightcliff foreshore, East Point and Buffalo Creek in Darwin.

Where to find The best places to look for this bird are Buffalo Creek, Lee Point and East Point in Darwin.

Australian Yellow White-eye

Zosterops luteus 4–5 in | 10–13 cm
(Canary (or Yellow) White-eye)

This species is mainly coastal, being found in mangroves and sometimes nearby monsoon forest or other dense vegetation. Usually found in pairs or small parties, their yellow colouration makes them difficult to see as they glean insects in the outer foliage of the mangroves. It is an attractive bright-yellow bird, slightly greenish above and with a conspicuous white eye-ring.

Red-headed Honeyeater

Myzomela erythrocephala
4–5 in | 11–13 cm

This pretty little bird is a common resident in mangroves and gallery forest in the northern part of the Top End. It is usually found close to water, particularly in mangroves and woodland close to waterways around the coast or along large rivers. Males are small and dark brown, with a long, downcurved bill and a bright-red head. Females are paler brown with a faint red wash around the face.

Where to find Anywhere there are mangroves; Buffalo Creek and East Point in Darwin are good places to look.

Gerygones are a genus of small birds, with one or two species present in most places throughout the Top End. They build fantastic, messy, hanging nests that often look like debris draped from a tree branch, usually over water, and are frequently parasitized by Little Bronze-cuckoos (*page 103*).

☐ Mangrove Gerygone

Gerygone levigaster 4½ in | 11 cm

Aptly named, this species is found exclusively in mangroves, usually close to the coast, and can be quite common. It looks very similar to Large-billed Gerygone, being greyish-brown above and white below, but has a small white eyebrow.

Where to find In mangroves anywhere around Darwin, including Leanyer sewage ponds and Buffalo Creek.

☐ Large-billed Gerygone

Gerygone magnirostris 4½ in | 12 cm

This species is most common in the northern Top End where it is often closely associated with dense waterside vegetation including monsoon forest and mangroves. A nondescript species, it is greyish-brown above and white below, with a distinct white eye-ring.

Where to find Howard Springs is a good place to look for this bird.

◀ ☐ **Green-backed Gerygone** *Gerygone chloronata* 4 in | 10 cm

This tiny bird is found in monsoon forest, lush parks and gardens, and woodland along waterways throughout the Top End. Since it spends most of its time high in the canopy, its repetitive, high-pitched, ratchetty call is the best way to find one. Given a good view, you will see its white underparts, olive-green wings and grey head.

Where to find East Point and Buffalo Creek in Darwin are good places to see this species.

☐ **Varied Triller** *Lalage leucomela* 7½ in | 19 cm

This species is fairly common in monsoon forest and lush parks and gardens across the northern Top End. A type of cuckooshrike (see *page 175*), it is usually seen foraging in the canopy where it can be quite difficult to see. Males are black above with white edging to many of the wing feathers, and fine black barring on the underparts. Females are similarly patterned but much duller, appearing more dusky than the black-and-white male.

Where to find Howard Springs, East Point and Buffalo Creek in Darwin are all good places to search.

Mangrove Golden Whistler

(Black-tailed Whistler)
Pachycephala melanura
6–7 in | 15–17 cm

Difficult to see in its dense habitat, this species is found mostly in large stands of mangroves, and occasionally in thick vegetation along some of the larger tidal rivers. A good technique for bringing birds closer, and a technique that works with many other species, is to make a loud squeaking sound by kissing the back of your hand. Making a loud "pishing" sound through clenched teeth can also work, the birds often coming to investigate. Males tend to be shyer than females, remaining wary and approaching furtively through thick cover. The males of this attractive species are easily identified, being bright yellow below and green above with a black head and white throat. However, females are a little more difficult, being much duller, with a greenish-brown back, pale throat and yellowish underparts.

Where to find Can be found in most large areas of mangroves close to Darwin, including Nightcliff and Buffalo Creek.

Where to find Good places to look around Darwin include East Point and Howard Springs.

☐ Grey Whistler

Pachycephala simplex

6 in | 15 cm

This very plain bird would be unlikely to get your attention were it not for its beautiful, lilting whistle – a frequently heard sound in monsoon forests, lush parks and gardens of the Top End. Usually seen feeding unobtrusively in the mid-storey, it is whitish below and brown above with a pale grey head.

☐ Little Shrike-thrush

Colluricincla harmonica

7½ in | 19 cm

Often found feeding in mixed flocks with other birds such as flycatchers or honeyeaters, this species is reasonably common across the Top End, and is usually found in dense vegetation such as monsoon forest, lush parks and gardens. A fairly nondescript small bird, it is brown above and buffish below with a dark bill.

Where to find Howard Springs is a good place to look.

Australian Figbird

Sphecotheres vielloti
10½–11½ in | 27–29 cm

Where to find Easy to see in many of the parks and gardens around Darwin.

A noisy and common resident of Top End towns, parks and gardens, this bird is often found in large groups, particularly around fruiting trees. The males are green above and yellow below, with a black head and bright-red patch of skin around the eye. Females are dark olive-green, with a pale, heavily streaked breast.

Green Oriole

Oriolus flavocinctus
(Yellow Oriole)
10–11½ in | 25–29 cm

Where to find Good places to search around Darwin include Buffalo Creek, East Point and Howard Springs.

Sounding like water glugging out of a bottle, the call of this bird is a familiar sound in parks and gardens throughout the Top End. Although called 'Yellow' Oriole in Australia, this name is shared with another bird, *Icterus nigrogularis*, found in South America. International birders therefore call the Australian species Green Oriole, perhaps a more accurate name as it is yellowish-green and covered in dark streaks.

☐ Northern Fantail

Rhipidura rufiventris

6½–7 in | 16–18½ cm

This small bird is common across the Top End, usually in denser vegetation such as monsoon forest and along rivers and creeks, and in parks and gardens. Unlike most other flycatchers and fantails, which are continually active, the Northern Fantail is usually found sitting quietly on an exposed branch in the forest, occasionally sallying out to catch insects. It is plain grey with an obvious white throat and buffish breast.

Where to find Easy to find around Darwin at East Point and Howard Springs, and throughout Kakadu NP.

☐ Arafura Fantail

Rhipidura dryas

6–6½ in | 15–16 cm

This fantail is seemingly in perpetual motion, continuously fanning its tail and flicking it from side to side. It is an uncommon species across the Top End, and most often found in monsoon forest or thick vegetation near water. Brown above and pale below with a white throat, it has a rufous rump and long, white-tipped tail.

Where to find Sometimes seen on the Yellow Water cruise in Kakadu NP, and also at Buffalo Creek and Lee Point in Darwin.

Spangled Drongo

Dicrurus bracteatus

11–12½ in | 28–32 cm

Common in wetter habitats such as monsoon forests and lush parks and gardens, these birds are readily told from other similar species by their strange 'fish-shaped' tail. At first sight they appear black, but if you get a good look at one in bright light, you should see the small, iridescent streaks covering the breast and head which give the species its name.

Where to find Easily seen around Darwin at sites like East Point and Howard Springs.

Black Butcherbird

Cracticus quoyi

15½–17 in | 39–43 cm

Common in monsoon forest, parks and gardens around Darwin, these birds are voracious predators, catching small mammals, lizards and birds. Like all butcherbirds they use their powerful hooked bill to dismember their prey, sometimes impaling it on a stick or wedging it in a forked branch as they do so. Adults are entirely black, while younger birds are brown.

Where to find Easily found at East Point and Buffalo Creek in Darwin.

Where to find Often seen on Yellow Water in Kakadu NP, and can also be found in mangroves around Darwin at sites like Leanyer sewage ponds.

☐ Broad-billed Flycatcher

Myiagra ruficollis

6–7 in | 15–17 cm

Despite its name, the bill of this flycatcher is not obviously broader than that of other similar species. It is common in mangroves and dense vegetation near water, mostly along the coast, but also around large wetlands and rivers in the northern Top End. Both male and female are very similar to the female Leaden Flycatcher (*page 184*), with a grey back, white breast and pale orange throat.

☐ Shining Flycatcher

Myiagra alecto

6–7 in | 15–18 cm

These vocal birds are usually seen in pairs, flying along the fringes of well-vegetated creeks and rivers, often low over the water. They build beautiful cup-shaped nests, tightly bound with cobwebs, usually over water. The sexes are very different, females being white below and rufous above with a blue-black cap, while the males are glossy blue-black all over.

Where to find Kakadu NP is a good place to look for this flycatcher; it is often seen along the river at Mardugal and on the Yellow Water cruise.

Where to find Kakadu NP is a good place to look, particularly around the edges of rivers and billabongs like Mardugal and Mamukala. Can also be found in Elsey NP near Mataranka and in Timber Creek.

☐ Buff-sided Robin *Poecilodryas cerviniventris* 6–6½ in | 15–17 cm

This widespread but uncommon species has very specific habitat preferences, being found only in thick monsoon or gallery forest vegetation along waterways and around wetlands. Shy but inquisitive, they are nearly always found in pairs, with each pair maintaining a territory along a stretch of creek or billabong. The birds spend most of their time flitting about in the undergrowth with their tails cocked, and are usually detected by their call, a loud "*ch-chp…cheew*!" or a piping three- to four-note whistle. This piping call is easily imitated and will often bring the birds closer to investigate. They are quite pretty birds and readily identified by the combination of dark brown upperparts, grey breast and buffish flanks, and the prominent white eyebrow and flash in the wing.

Where to find Can be found in a few places around Darwin, including the mangroves around the sewage ponds at Leanyer and Palmerston.

☐ Mangrove Robin *Eopsaltria pulverulenta* 6–6½ in | 15–16 cm

This inconspicuous robin is fairly common in mangroves around the coast, and along some larger tidal rivers. It often occurs in pairs, which spend most of their time, deep in the gloom of mangrove forests, feeding among the complex root systems and low foliage, and occasionally dropping down to the mud to pick up food. Its presence is most often detected by the distinctive two-note whistle. Although shy, these birds are inquisitive and will sometimes come to investigate an imitation of their call. A technique birders use to attract birds, which works well with this species, is 'pishing' – saying "*psshhh*" repeatedly through clenched teeth will usually bring in smaller birds to investigate. Mangrove Robins are easily identified, being steely grey above and white below.

Birds of Open Areas
Woodlands and Grasslands

This section covers both the open tropical woodlands which dominate the Top End's landscape, and the open grasslands that are often found in association with them. These open grassy woodlands are the most widespread habitat in the Top End, being found from the northern coasts, south to the southern edge of the region and beyond. Unsurprisingly, this habitat holds many of the Top End's bird species, including some of Australia's most sought-after birds, such as Gouldian Finch and Hooded Parrot.

Variegated Fairywren

Gouldian Finch

Red-collared Lorikeet

Barking Owl

☐ Australian Bustard *Ardeotis australis* 31½–47 in | 80–119 cm

Northern Australia, and particularly the Top End, is one of the strongholds for these unusual birds which are now scarce across much of Australia, having been a popular target for hunters. They are fairly common in drier regions, where they can be found wandering in loose groups, always with their head held high and pointing up at a 45-degree angle. They prefer open areas, and are often seen on airfields and farmland, or in open savanna woodland. During the breeding season males form 'leks' or groups that may be spread over a wide area, with the males up to several hundred metres from each other. Each male performs a spectacular display, extending a large, dangling throat sac and raising his tail over his back before emitting a strange roaring call. Females move around the lek inspecting the males before choosing their favourite for mating.

Where to find There are usually a few birds on the airfield at Katherine, and they are commonly seen wandering through the open savannas between Katherine and Timber Creek.

☐ **Bush Stone-curlew** *Burhinus grallarius* 21–23 in | 53–58 cm
(Bush Thick-knee)

These cryptically patterned birds, a type of wader, are a familiar species across the Top End, and responsible for the wailing "*weee-loo*", "*weee-loo*" so often heard at night. This haunting call is regularly given just on dusk, and often in chorus by a pair or small group of birds.
They are widespread and quite common, usually occurring in open, grassy areas with scattered trees, and many suburban parks and playing fields host a resident pair. As the very large yellow eyes suggest, Bush Stone-curlews are mostly nocturnal, emerging after dark to feed on insects and other invertebrates. They are quite common in urban areas and are often seen crossing streets at night on their long spindly legs, or feeding on insects attracted to street lamps.

Where to find In parks and gardens throughout Darwin; East Point Reserve is a good place to look.

During the day these birds sit or stand quietly among leaf-litter, often in the shade of a tree or bush, where they are well camouflaged and very difficult to see. In the dry-season they sometimes occur in quite large groups, but at other times are found in pairs. They nest on the ground, usually under or near some cover and while one adult incubates the eggs, the other will defend their territory, fearlessly approaching intruders and hissing with wings spread and tail cocked. The tiny, fluffy chicks are precocious, and able to run around following their parents as soon as they hatch.

Pheasant Coucal *Centropus phasianinus* 21½–27 in | 55–69 cm

Although secretive, these comical birds are still a common sight across the Top End, and may be seen anywhere there is grassland or scrubby thickets. A large, scruffy-looking bird with a long, ragged tail, they are often seen running across the road, or sometimes flying from one patch of cover to the next. They are poor fliers, flapping frenetically as they steadily lose height before crash-landing at their destination. Perhaps the best indication of their presence is the unusual call, which once learned is heard quite often – it sounds like water glugging out of a bottle in the distance. A good time to keep an eye out for these birds is after a shower of rain, when they will often clamber up onto an open branch, holding their wings half-open to dry off.

Breeding

Non-breeding

Where to find Good places to look include the grassy margins of Yellow Water in Kakadu NP, or in open grassy woodlands.

Where to find Occurs anywhere across the Top End where there is dense grass.

■ Brown Quail *Coturnix ypsilophora* 6½–8 in | 16–20 cm

Like all quail, Brown Quail are ground-dwellers, spending most of their time in dense grass. Often the only clue to their presence is a drawn-out two-note whistle "*too-weeee*", with a rising inflection on the last note. This species usually occurs in small parties called coveys, and although difficult to observe, with patience you may see them feeding quietly by the side of a trail or road, or running away in front of you with their necks held forward. The best time to look for them is early in the morning or late in the afternoon, and they will visit waterholes to drink late in the dry-season. They are more often flushed unexpectedly from long grass beside the trail, leaping high into the air and rocketing away low over the grass before crash-landing into cover. Their landing technique is curious; rather than diving back into the grass like some other quails, they seem to land bottom first, with the body held vertically. Brown Quail are cryptically patterned like most other quail, with plump bodies and small heads. The body and wings are brown and covered in fine black streaking and barring.

☐ **Australian Koel** *Eudynamys cyanocephalus* 16 in | 41 cm

(Eastern Koel; 'Rainbird'; 'Stormbird')

This large, long-tailed cuckoo is a migratory species that arrives in the Top End from New Guinea around September. The male's continuous "*ko-el, ko-el, ko-el*" call is a sure sign of both its arrival and the impending wet-season (hence their alternative names). Although noisy, koels are quite shy, with the male often calling from the cover of a tall, dense tree, making him difficult to see well. An imitation of his call rarely brings him out into the open, but seems to send him into a frenzy of louder and more frenetic calling. The cryptically patterned female lays her eggs in the nests of larger birds, including Blue-faced Honeyeaters (*page 168*), friarbirds (*pages 168–169*) and Magpie-larks (*page 183*).

Where to find Common in wooded habitats, particularly in suburban parks and gardens from Darwin to Mataranka.

☐ Channel-billed Cuckoo *Scythrops novaehollandiae* 22½–27½ in | 56–70 cm

Like the Australian Koel, these enormous cuckoos arrive in Australia from New Guinea to breed around September, with their raucous "*krrorrk, krrorrk, krrorrk*" calls announcing their arrival. Although uncommon in the region, they occur in most wooded habitats, particularly along rivers, and are most often seen flying overhead and calling, or hanging around fig trees in small parties of two or three birds. Like all cuckoos they are brood parasites, laying their eggs in the nests of larger birds; in the Top End, Torresian Crows (*page 186*) are the primary host. Channel-billed Cuckoos are unmistakable grey birds with slightly darker, black-scalloped wings and a massive horn-coloured bill. In flight the long, pointed wings and long tail give them a distinctive silhouette, like a huge flying 'T'.

Where to find There are often a few pairs in the woodland around Bitter Springs near Mataranka or in Kakadu NP.

Where to find Quite common in towns including Pine Creek, Katherine and Mataranka.

☐ Crested Pigeon *Ocyphaps lophotes* 12–14 in | 30–36 cm

This bird is sometimes mistakenly called 'Topknot Pigeon' due to the long, black crest that sticks up from the top of the head, but that is the name given to another species (*Lopholaimus antarcticus*) which is found in the rainforests of eastern Australia. Usually seen foraging quietly on the ground in town parks and gardens, Crested Pigeons are found throughout the Top End but are most common in the drier parts of the region, south of Pine Creek. When flushed they tend to 'burst' off the ground, with the wings making a strange whistling sound, and birds always tip their tail up as they land. They are soft bluish-grey and have black barring on the shoulders and a panel of iridescent feathers on the wings that glisten when the light catches them at the right angle. With good views you may also see a patch of soft pink on the side of the neck.

☐ Spinifex Pigeon

(White-bellied Spinifex Pigeon)
Geophaps plumifera
7½–9 in | 19–23 cm

Where to find Never easy to find, the best places to look include Joe Creek Campground near Victoria River Crossing and Policeman's Point in Timber Creek.

These small, dumpy pigeons are a prize sighting in the southwestern Top End. They prefer rocky areas and are usually seen scurrying around on the ground or feeding beside roads, their rusty-red plumage blending in well with the red rocks and soil. A good strategy for finding them, particularly late in the dry-season, is to watch for birds coming to drink at small waterholes towards dusk.

☐ Common Bronzewing

Phaps chalcoptera 12–14 in | 30–36 cm

Fairly common in the drier southern areas of the Top End, these large, plump pigeons are often seen feeding quietly by the roadside. They appear dull brown from a distance, but up-close are rosy-brown below, with beautiful iridescent feathers on the wings, and a cream forehead.

Spinifex Pigeon

Common Bronzewing

Where to find Seen regularly around Katherine and also in Elsey NP near Mataranka.

☐ Chestnut-quilled Rock-pigeon *Petrophassa rufipennis* 11–12 in | 28–31 cm

One of only a handful of birds that are found only in the Top End, this species occurs exclusively around the rocky escarpments of Arnhem Land. Although very restricted in distribution by its habitat preference, this pigeon can be quite common in some areas. It is a shy species, usually flying off with a clatter at the first sign of disturbance – the only opportunity you will have to see the chestnut patches in the wings. This behaviour, and the well-camouflaged plumage make it a difficult bird to see well; a good strategy is to sit quietly in a spot where you can watch over a large area of escarpment and look for birds resting or scurrying around on rocky ledges or among boulders. The similar **White-quilled Rock-pigeon** *Petrophassa albipennis* (not illustrated) has white wing patches and is found in the far southwest of the Top End.

Where to find Kakadu NP is the best place to find this pigeon. Search for them at Ubirr, or take the hike to the top of Gunlom and search the escarpments there. The White-quilled Rock-pigeon can be found on the escarpment walk at Victoria River Crossing.

☐ **Partridge Pigeon** *Geophaps smithii* 10–11 in | 25–28 cm

These medium-sized ground-dwelling pigeons were once common across the Top End, but have become quite difficult to find since the turn of the century. They are usually seen in small groups foraging on open ground, often around the edges of picnic areas or campgrounds or beside quiet roads. When disturbed, these birds tend to freeze, making them very difficult to see, and if approached more closely will run away, keeping their brown backs towards the observer. If followed they will eventually flush, exploding up off the ground with a clatter of wings and flying away quickly in powerful, direct flight. If you happen to flush birds unwittingly, stop and look closely around on the ground nearby and you may find other birds standing still, quietly watching you. If you are lucky enough to get good views of one, look for the patch of bright red skin around the eye, the white breast, and a panel of iridescent purple feathers in the wing.

Where to find The best place to see this bird is around Pine Creek, including Copperfield Dam and Umbrawarra Gorge.

☐ **Diamond Dove** *Geopelia cuneata* 7½–9 in | 19–23 cm

Australia's smallest dove, this species is usually seen feeding quietly on the ground, particularly around picnic areas, parks and gardens, often in mixed groups with the similar Peaceful Dove. Found throughout the Top End, like many of the dry country birds in the region its numbers fluctuate according to the conditions. It is most common in the drier woodlands around Katherine and southwest to Timber Creek, where some birds are present year-round. In the dry-season it may be found more regularly near the coast, sometimes even around Darwin. It has a plain blue-grey head and breast, white belly and brown wings peppered with tiny white spots, or 'diamonds'. At close range you can also see the diagnostic red eye-ring.

Where to find The grounds of the Victoria River Roadhouse and anywhere around Timber Creek are good places to look.

☐ **Bar-shouldered Dove** *Geopelia humeralis* 10–11½ in | 25–29 cm

Found in most wooded habitats across the Top End, this medium-sized dove has adapted well to human settlement and is common in suburban parks and gardens. It is usually seen feeding quietly on the ground in pairs or small groups, and sometimes in mixed groups with Peaceful Doves. It tends to stay close to cover and is rarely found far from water. A beautiful bird, it has a blue-grey head and breast, large rufous neck patch, brown wings and white belly with a long brown tail. The top of the head, back of the neck and the wings are covered in fine black scalloping. It has a distinctive rufous patch in the outerwing, which is a good feature to look for as it flushes from the ground.

Where to find Easily seen in town parks and gardens.

☐ **Peaceful Dove** *Geopelia placida* 7½–9 in | 19–23 cm

The often-repeated, soft "*doodle-doo*" of these lovely little doves is a constant part of the Top End's soundtrack and, once learned, it seems there is nearly always a bird within earshot. Found in most open habitats across the Top End, they are unobtrusive birds and spend much of their time feeding on the ground, where they are usually seen scurrying around in pairs or small parties. If disturbed they will flush with a flurry of wingbeats, usually perching in a nearby tree and nervously watching on before returning to the ground to continue feeding. They are beautifully patterned soft grey and brown, with fine black scallops on the back and head. They also have a ring of pale blue skin around the eye which is visible from close range.

Where to find Common in suburban parks and gardens throughout the region.

Diamond Dove

Bar-shouldered Dove

Peaceful Dove

Barking Owl *Ninox connivens* 15½–17½ in | 39–44 cm

This owl is common in woodland, monsoon forest and parks across the Top End, particularly near watercourses. It is best known for its distinctive "*woof-woof*" call, which sounds exactly like a barking dog. Females also give a blood-curdling, falsetto scream when preparing to breed, a very disconcerting sound to hear in the middle of the night! Often found in pairs, Barking Owls spend the day roosting quietly in leafy trees. As dusk falls the male and female begin calling together before leaving their roost to hunt for mammals and birds, often returning to the same roost the following day. A large owl, it is brown above and boldly streaked below, with large yellow eyes and powerful yellow talons.

Where to find Frequently encountered in the suburbs of Darwin; another good location is the grounds of the Victoria River Roadhouse.

Southern Boobook *Ninox boobook* 12–14 in | 30–36 cm

With a variety of colloquial names across the country, including 'Morepork' and 'Mopoke', this is perhaps Australia's best-known owl, and its distinctive high-pitched "*boo-book*" or "*more-pork*" call is a frequent sound at night in the Australian bush. It is common across the Top End in almost any wooded habitat, from monsoon forest to dry, open woodlands, parks and gardens. Like most owls they roost in thick vegetation, tree hollows and sometimes on sheltered rock ledges during the day, emerging to hunt insects and small vertebrates at night. This species is often confused with the more uniformly coloured Barking Owl, which is much larger and more powerful, and has large yellow eyes and huge talons. By comparison, Southern Boobook is a small owl, brown above and streaked below; the wings are often spotted with white and the face has distinctive brown 'goggles' around the eyes.

Where to find Virtually anywhere there are trees.

Eastern Barn-owl *Tyto (alba) delicatula* 11½–15 in | 29–38 cm

(Common Barn-owl)

Barn-owls are familiar to most people as one of the owls of Harry Potter, nursery rhymes and other stories. Their ghostly white shape is most often seen at night, sitting on a roadside fence post or sailing low over a field on upswept wings, searching for prey. They roost in tree hollows, caves or abandoned buildings, emerging at night to hunt for small rodents. Although they are not as common in the Top End as elsewhere in Australia, this owl's numbers can vary wildly in response to rodent plagues or similar events.

Where to find Can turn up almost anywhere, but the best places to look are around open fields and grasslands.

Tawny Frogmouth *Podargus strigoides* 14–20 in | 36–51 cm

Frogmouths spend their day sitting quietly in a tree, and are best known for their habit, when disturbed, of freezing, slowly extending their bill skywards, flattening their feathers close to their body, and doing an impressive imitation of a broken branch. If you find one in this pose look closely; although its eyes appear closed, they are actually slightly open, and it will be watching you intently! Tawny Frogmouths are often found in pairs, so if you find one bird, search the surrounding trees for its mate. At night they hunt for insects and other small prey, often sitting quietly on a low branch, overhead power line, even on the road. They seem unafraid at night, and can be approached quite closely. Although not often seen during the day they are surprisingly common, with many suburban parks having a resident pair.

Where to find Almost anywhere there are trees.

■ Australian Owlet-nightjar *Aegotheles cristatus* 8½–9½ in | 22–24 cm

Closely related to nightjars, which are large birds that spend much of their time resting on the ground, owlet-nightjars behave more like small owls, hence the name. They are lovely little birds but are rarely seen during the day, as they spend most of their time sheltering in small holes in trees. They sometimes shuffle to the entrance of their hollow in the early morning to soak up the sun, but otherwise stay hidden from sight. At night they emerge to hunt actively for small insects, and can be quite vocal, making a variety of squeaking and churring calls that are commonly heard at night in campgrounds across the Top End. Owlet-nightjars can be difficult to find at night, because although they have large eyes, they have only faint eye-shine which can be difficult to detect with a spotlight.

Where to find Can be found almost anywhere there are trees.

Spotted Nightjar *Eurostopodus argus* 12 in | 30 cm

This nocturnal species is found across the Top End but is most common in drier woodlands away from the coast, particularly around stony ridges. It is rarely seen during the day, relying on its excellent camouflage to avoid predators as it sits quietly on the ground among leaf-litter, only flushing when about to be trodden on. Dusk is a good time to see this bird since it usually emerges before it is completely dark, flying fast on long wings, fluttering and gliding acrobatically, as it hunts for insects in open areas, along roads or over waterholes. It often sits on quiet roads at night and this habit provides the best way to see it, as the red eye-shine can be seen from a great distance in the car's headlights. If you spot one this way, slow down and approach slowly; usually it will sit tight on the road, and you can sometimes even get out of the car for a closer view. This affords a great opportunity to admire the intricately patterned plumage. In flight, this nightjar has conspicuous white wing patches that also show up in headlights or a spotlight beam. You may also come across juvenile birds, which look quite different from the adults, being rufous-brown all over. Unlike the similar but smaller Large-tailed Nightjar (*page 105*), the Spotted Nightjar does not have any white in the tail; however, these two species generally occupy quite different habitats.

Where to find Along roads at night in the southwest of the region, particularly around Victoria River Crossing and Timber Creek.

☐ Blue-winged Kookaburra *Dacelo leachii* 15½ in | 39 cm

Kookaburra is the name given to a genus of large kingfishers found in Australia and New Guinea, with the **Laughing Kookaburra** *Dacelo novaeguineae* of eastern Australia (not illustrated) being one of Australia's best-known birds. The species found in this region is the Blue-winged Kookaburra, and its raucous, maniacal, some would say unpleasant, call is a frequent sound of the Top End's savannas. It is often given in concert by small parties of birds that seem to be egging each other on. Despite the name, this bird is often found far from water and will eat almost anything it can catch, from small mammals to lizards, snakes, insects and even nestling birds. This species has a pale head and large blue wing patch, with males having a blue tail and females a brown tail with black barring.

Where to find Most campgrounds in Kakadu NP will have a resident pair.

☐ Forest Kingfisher

Todiramphus macleayii
7½–8½ in | 19–22 cm

This stunning kingfisher is a common sight in most wooded habitats throughout the Top End. It is quite a vocal species, giving a variety of harsh chattering and scolding calls, and is often seen in pairs, perched in the open on power lines or trees. If seen well in full sun, the sky-blue back, dark blue head and white breast is an unforgettable sight. In flight, a distinctive white window is visible in the wing.

Where to find Often seen on the Yellow Water cruise in Kakadu NP.

☐ Sacred Kingfisher

Todiramphus sanctus
8½ in | 22 cm

Although this kingfisher is present year-round in the Top End, numbers increase in the middle of the year when migrants from southern states arrive to spend the dry-season in northern Australia. It is common in most habitats across the Top End, and is not necessarily always found close to water. It is often confused with Forest Kingfisher, which has a pure white breast and is blue above rather than greenish-blue.

Where to find Can be found almost anywhere there are trees.

Forest Kingfisher

Sacred Kingfisher

Where to find Can be seen almost anywhere, often in suburban parks and gardens.

☐ Rainbow Bee-eater *Merops ornatus* 8½–10 in | 22–25 cm

These stunning and unmistakable birds are common in most habitats across the Top End, their melodic, telephone-like "*prrrp-prrrp*" calls being a familiar sound. Although there is a small resident population in the region, a significant influx of birds occurs in the dry-season when many migrate north from the southern states. They are most often seen around open areas, spending much of their time hawking gracefully for insects which are taken back to a perch and beaten against a branch before being eaten. They nest at the end of a narrow tunnel that they dig into a sandbank, making a small chamber at the end where they lay their eggs and raise their chicks.

Where to find Almost anywhere, particularly in suburban areas and around waterways.

☐ (Oriental) Dollarbird *Eurystomus orientalis* 10–11½ in | 25–29 cm

So called for the white 'dollars' in the outer part of their wings in flight, Dollarbirds are breeding migrants to the Top End from New Guinea, arriving around September and departing in April. After arriving they are noisy and prominent, giving their harsh "*kek-kek-kek*" calls as they perform acrobatic display flights on their long wings. Otherwise, they spend most of their time sitting quietly on a favourite exposed perch, sallying out occasionally to pursue insects. They are beautiful birds readily identified by their a greenish body, dark grey head, purple throat and red bill.

Red-tailed Black-cockatoo

Calyptorhnychus banksii
21½–23½ in | 55–60 cm

Watching a flock of these cockatoos with their slow, lumbering flight, the red tail panels of the males catching the sun as they wheel in to land at a small waterhole, is one of the most memorable sights of the Top End. Even though they are so large, these cockatoos can be quite unobtrusive; the key to finding them is listening for their soft, grating call which sounds like a squeaky gate. They are often seen feeding on the ground in burnt areas, when they can be quite approachable. Males are plain black with red panels in the tail, while females are covered in tiny yellow spots and have barred orange panels in the tail.

Where to find Quite common across the Top End and seen regularly in suburban areas from Darwin to Katherine.

■ **Cockatiel** *Nymphicus hollandicus*
11½–12½ in | 29–32 cm

A well-known cage bird throughout the world, these rather dowdy little parrots make for a beautiful sight when seen streaking across the sky in small, tight flocks, twisting and turning on lazy wing beats and giving their bubbly, melodic call. Common across inland Australia, Cockatiels reach the northern limit of their distribution in the southern Top End. Although birds occasionally reach the outskirts of Darwin they are most numerous in the drier savannas from Pine Creek south to Mataranka and southwest to Timber Creek. Males are grey with a yellow head and orange ear patch, while females are similar but duller, with faint barring on the wings and tail.

Where to find Copperfield Dam near Pine Creek is a good place to search. Otherwise, keep an eye out for them on the highway between Katherine and Timber Creek.

Sulphur-crested Cockatoo

Cacatua galerita

19–21½ in | 48–55 cm

Australia's famous 'white' cockatoo, these birds are particularly common in suburban areas where there are plenty of trees on which they can feed, and where they often associate with Little Corellas and Galahs. The distinctive beautiful yellow crest is usually fanned when the bird lands, or is agitated.

Where to find Frequently seen in parks and gardens.

Little Corella *Cacatua sanguinea*

14–15½ in | 36–39 cm

Usually seen in large, noisy flocks, this species is particularly common in the Top End. It is a regular sight around Darwin and the larger towns, including Katherine and Mataranka, where it is often seen in mixed flocks with Galahs and Sulphur-crested Cockatoos. A small cockatoo, it has a small crest and a patch of bare blue skin around the eye.

Where to find Easily seen in suburban parks in Darwin and Katherine.

Galah *Eolophus roseicapilla* 14 in | 36 cm

This well-known Australian cockatoo is surely one of Australia's most beautiful birds, with flocks twisting and turning a cross the sky being a common sight throughout the country. The Galah is perhaps best known for its seemingly crazy behaviour – often hanging acrobatically from branches or power lines, or chasing one another for no obvious reason except that it seems like fun; in fact Australians often use the term 'galah' to describe someone who is acting like a clown. It is easily identified, having a rose-pink head and body, white crest and grey wings.

It is worth noting that the correct pronunciation of this unusual name is *"g-LAH"*, not *"GAY-la"*.

Where to find Can be found just about anywhere, but particularly in drier areas around Katherine.

☐ Red-collared Lorikeet *Trichoglossus rubritorquis* 10½–12 in | 26–30 cm

(Rainbow Lorikeet)

If this is not Australia's most beautiful bird, it would certainly be near the top of the list. Visiting birders often call Australia the 'Land of Parrots', and when you realise that most of these people come from countries where parrots do not occur, it is easy to see why. We certainly take for granted the fact that such a stunning species is actually a common garden bird. Red-collared Lorikeets often congregate in large, noisy flocks around flowering eucalypts, where the birds use their specially adapted brush-tipped tongue to extract nectar from the blossoms, often squabbling with each other and aggressively chasing other species away from 'their' patch. Unlikely to be confused with any other bird, this lorikeet has a blue head with fine, lighter blue streaks. The wings and tail are green and it has an orange breast and collar.

Where to find Easily found in suburban areas including Darwin and Katherine.

☐ Varied Lorikeet *Psitteuteles versicolor* 7–8 in | 18–20 cm

Like many Australian birds, particularly those that rely on flowering trees as their primary food source, Varied Lorikeets are nomadic, moving into and out of an area depending on whether there are flowering trees around. There are usually small numbers present in the savanna woodlands throughout the Top End, but they can be difficult to find, and are most often seen flying around in small flocks in the early morning or late afternoon. The key to getting good views of this beautiful little parrot is finding a patch of flowering eucalypts. There may be hundreds of birds in the vicinity, constantly giving their high-pitched screeches and zipping around between trees or clambering nimbly through the foliage gorging on nectar. A stunning species, these lorikeets have green bodies with a pale pink and purple neck and a red cap. With good views you may see the broad white eye-ring and fine yellow streaks which cover the entire body.

Where to find May occur anywhere, but there are often a few birds around the cemetery at Adelaide River, and around Katherine and Timber Creek.

☐ **Northern Rosella** *Platycercus venustus* 12–12½ in | 30–32 cm

These beautiful parrots are the Top End's representative of a group of parrots that is widespread throughout most of Australia and since this species is only found in northern Australia, it is particularly sought after by visiting birders. Most common in the central Top End, they can sometimes be difficult to find, being inexplicably absent from apparently suitable habitat. They are easily identified, having a black head and white cheek patch, a pale straw-coloured body covered in black scalloping, and blue wings and tail.

Where to find Good places to search include Copperfield Dam near Pine Creek and the campground at Gunlom in Kakadu NP.

☐ **Red-winged Parrot** *Aprosmictus erythropterus* 12–12½ in | 30–32 cm

These stunning parrots look almost out of place in the Australian bush, the dazzling green with red flashes contrasting with the dull greens, reds and greys of the landscape. They are usually seen in pairs or small parties, and their rapid flight with regular, deep wing beats is distinctive. They are common across northern Australia, particularly in the Top End where they are often found in towns feeding on suburban fruit trees, or drinking at waterholes. The males are more brightly coloured than females, which are dull green all over with only a small red patch on the wing.

Where to find Can be seen in Darwin, but are particularly common around Pine Creek and Katherine.

Where to find The best place to look is Pine Creek, where they can be found around the township feeding on lawns, or at Copperfield Dam. They can sometimes also be found around Katherine and along the road from the main highway to Edith Falls.

☐ Hooded Parrot *Psephotus dissimilis* 10–11 in | 25–28 cm

Quite rare, these lovely little parrots are found only in the savanna woodlands of the central Top End, and nowhere else in the world. An unusual species, Hooded Parrots excavate tunnels in large terrestrial termite mounds, making a chamber at the end in which they lay their eggs and raise their chicks. They spend most of their time feeding quietly on the ground in small flocks consisting of a few males, several females, and often a number of younger birds, which have similar plumage to the females. If you flush them from the ground they will usually fly up into nearby trees, watching you nervously, but if you are patient will usually return to the ground to continue feeding. Another good strategy to see them is to wait quietly by a waterhole late in the dry-season, when the birds will often come to drink at dawn and dusk. Males are a beautiful turquoise-green below and have a black cap, brown wings and a large golden patch on each shoulder, while females are dull grass-green. In flight they have a distinctive silhouette, with a small body, longish tail and short, broad wings. Their flight is fast and undulating.

☐ Great Bowerbird *Chlamydera nuchalis* 12½–14½ in | 32–37 cm

Bowerbirds are best known for their elaborate bowers, a construction of sticks and twigs built by the male solely to impress prospective mates. Some bowerbirds in the rainforests of New Guinea and eastern Australia are gaudily patterned, but these species build only rudimentary bowers. Generally, the duller the species, the more elaborate the bower – and as one of the dullest species, the Great Bowerbird's bower is a spectacular construction. It is an avenue up to a metre long of tightly packed twigs and small sticks, usually built in a sheltered spot beneath a bush or shady tree. Once complete, it is decorated with ornaments, often green or white items, such as bone, shells, pebbles, flowers, plastic bottle caps, glass, or just about anything the bird can find. During the breeding season, the male returns to his bower regularly throughout the day to tidy up or rearrange things, also giving his harsh call to try and attract a female. At the same time, he must defend his bower against other males, who will eagerly steal the best trinkets! A male will mate with as many females as he can attract to his bower; each female then builds a small nest and raises the chicks entirely on her own. A large, lanky bird with a stout black bill, Great Bowerbirds are dull brown with some darker brown patterning on the wings. Males have a pink crest on the back of the head that is usually hidden and only likely to be seen when the bird is displaying at his bower.

Where to find Easily seen around Pine Creek and Katherine, and there is often a bower in the grounds at the Victoria River Roadhouse which the staff may be able to show you.

Where to find The Arnhem Land escarpment subspecies can be found on any of the escarpments in Kakadu NP. The other subspecies can be found around Victoria River and Timber Creek.

☐ Variegated Fairywren *Malurus lamberti* 4½–5½in | 11–14cm

Like other fairywrens, this species occurs in small parties, usually with one or two stunning full-coloured males. Two subspecies occur in the Top End: one (subspecies *dulcis*) on the Arnhem Land escarpments that is endemic to the Top End and has bluish females; and another (subspecies *assimilis*) in the drier southwestern Top End, where the females are plain brown.

Male

Female *dulcis*

Where to find There are a couple of particularly good spots to look: the old bridge and boat ramp at Victoria River Crossing; and anywhere along the river at Policeman's Point in Timber Creek.

☐ Purple-crowned Fairywren

Malurus coronatus

5½–6 in | 14–15 cm

Found in the far southwest of the Top End, this beautiful fairywren has very specific habitat preferences, only occurring in thick grassy areas along riverbanks. Although its range is fairly restricted, it is actually quite common in suitable habitat, occurring in pairs and small family parties that are often heard chattering to each other. Only the male has a purple crown; females have a chestnut ear-patch.

Where to find Can be found anywhere there is rank grassland. Holmes Jungle in Darwin is a good place to look.

☐ Red-backed Fairywren

Malurus melanocephalus

4–5 in | 10–13 cm

The common fairywren across the Top End, this bird is easily seen at many locations. It occurs in small family parties, usually with one or two fully coloured males and a number of 'brown' birds which are either females or young males. All members of the group play a part in the breeding duties, building nests and raising the young. The males are very attractive little birds, with a jet black body and red back.

Purple-crowned Fairywren

Red-backed Fairywren

Where to find Although difficult to find, they are not uncommon in some remote parts of the escarpment in Kakadu NP, but unfortunately these areas are generally inaccessible, requiring special permits and long, multi-day hikes or helicopter flights. There used to be a population at the top of the waterfall above Gunlom in southern Kakadu NP, but they seem to have disappeared from that site after recent fires.

☐ White-throated Grasswren *Amytornis woodwardi* 8–9 in | 20–23 cm

These stunning but shy little birds are found nowhere else in the world except the rugged spinifex-clad sandstone escarpments of western Arnhem Land. After being discovered by Europeans early in the 20th century, their inaccessible habitat resulted in them only being seen once more before they were found again in the late-1960s. Their rarity and the difficulty in seeing them, make them a major target for birdwatchers visiting the Top End. Living in small family parties, their secretive behaviour makes them very difficult to find, and they are usually only glimpsed fleetingly, scurrying around like small feathered rats, ducking behind clumps of spinifex or disappearing over rock ledges. The birds require mature spinifex to breed, as they build their nests deep in the spiky interior of these hummocks. Spinifex burns very easily, making the grasswrens very susceptible to fire, and there are suggestions that poor management of fire frequency in their habitat could be resulting in population declines. These long-tailed birds are readily identified by their chestnut body, black upperparts and breast covered in fine white streaks, and white throat.

Honeyeaters are a family of birds that is widespread in Australia, with many species common in cities and towns, and parks and gardens across the country. Most species feed primarily on nectar, which means they must move around in search of flowering trees. When there is a large flowering event, birds may be present in huge numbers, with most moving on once the flowering is over.

☐ **Banded Honeyeater** *Cissomela pectoralis* 4½–5½ in | 11½–14 cm

These small honeyeaters are quite common, with small numbers usually present in savanna woodlands throughout the Top End. Since they are blossom nomads, when a flowering event occurs numbers often increase rapidly, with hundreds of birds suddenly appearing overnight – at which time they may become the most common bird in an area. Easily identified, they are black above and white below, with a long, curved bill and black breast band. Younger birds have brown instead of black plumage.

Where to find Easily found in drier areas, and often seen in Kakadu NP or around Katherine.

☐ **Brown Honeyeater** *Lichmera indistincta* 5–6½ in | 12–16 cm

One of the Top End's most regularly seen birds, this honeyeater is found across the region and in most habitats. It is quite common in parks and gardens, and most backyards will have a resident pair which will vigorously defend any flowering plants or shrubs in their territory. A little difficult to identify because it is so plain, this small bird is brown above and paler brown below. The wings feathers have pale yellow edges and there is a diagnostic small yellow triangle behind the eye.

Where to find Easily seen in suburban parks and gardens.

☐ **Dusky Honeyeater** *Myzomela obscura* 5–6 in | 13–15 cm

(Dusky Myzomela)

Fairly common across the Top End in most habitats, including monsoon forest, woodland close to waterways, parks and gardens, this small, dark, chocolate-brown honeyeater has a longish, downcurved bill. It can be difficult to separate from Brown Honeyeater, but has a longer bill and is more uniformly dark brown; it also lacks the yellowish edges to the wing and the yellow triangle behind the eye of that species.

Where to find Easily found in parks and gardens around Darwin.

Banded
Honeyeater
Brown
Honeyeater
Dusky
Honeyeater

Each of these species is common right across the Top End, regularly being found in many town parks and gardens. They are often noisy and aggressive, spending much of their time flitting around in flowering trees, and chasing each other, or any other birds close by.

Where to find Common in woodlands around Katherine and Mataranka.

☐ Rufous-throated Honeyeater

Conopophila rufogularis

5½ in | 14 cm

This species is widespread across the Top End, but most common away from the coast, particularly in the drier savannas in the south of the region. It is similar to the Rufous-banded Honeyeater, but lacks the breast band and has a browner head and rufous throat patch.

Where to find Easily found in parks and gardens around Darwin.

☐ Rufous-banded Honeyeater

Conopophila albogularis

4½–5½ in | 12–14 cm

Can be found in most habitats across the Top End, but is particularly common in coastal areas. It is a small honeyeater, brown above and pale below with a greyish head and broad rufous band across the breast.

☐ Yellow-tinted Honeyeater

Lichenostomus flavescens

6 in | 15 cm

This species is fairly common in woodlands, parks and gardens in the southern Top End, and moves around in response to flowering events. The similar but much rarer **Grey-fronted Honeyeater** *Lichenostomus plumulus* (not illustrated) is sometimes found in the dry southwest around Timber Creek, but that species is duller greyish-green above and has a faintly streaked breast.

Where to find Common in wooded areas near Katherine, Mataranka and Timber Creek.

☐ White-gaped Honeyeater

Lichenostomus unicolors

7–8 in | 18–20 cm

One of the most widespread and most regularly seen birds in the Top End, this species is found in all habitats right across the region. A plain, grey-brown honeyeater, it has an obvious white spot near the base of the bill which is diagnostic.

Where to find Common in parks and gardens in Darwin.

Yellow-tinted Honeyeater

Rufous-banded Honeyeater
Rufous-throated Honeyeater
White-gaped Honeyeater

Except for the White-throated Honeyeater, which is widespread and common, the other species on this page have either quite particular habitat preferences, or a restricted range within the Top End. They are often targeted by visiting birders and with a bit of effort are fairly easy to find.

Where to find Good places to look are around billabongs in Kakadu NP, and Copperfield Dam near Pine Creek.

☐ Bar-breasted Honeyeater

Ramsayornis fasciatus

5½ in | 14 cm

This honeyeater is widespread across the Top End, but is nowhere common. It is usually found near water, particularly in paperbarks around the edge of wetlands, and in woodland close to rivers or creeks. It is a fairly plain bird, but has distinctive black barring on its white breast.

Where to find Can usually be seen around the rock art galleries at Nourlangie Rock.

☐ White-lined Honeyeater

Meliphaga albilineata

7½ in | 19 cm

One of only a few birds found in the Top End and nowhere else in the world, these honeyeaters are restricted to the sandstone escarpments of Arnhem Land where their distinctive melodic call is a familiar sound. They are quite plain, but have a fine white stripe from the base of the bill to below the eye.

Where to find Common in woodlands around Katherine and Mataranka.

☐ White-throated Honeyeater

Melithreptus albogularis
4½–5½in | 12–14cm

This small honeyeater is common in most wooded habitats across the Top End and is a regular visitor to parks and gardens. It is usually seen foraging actively in small parties or flocks in the outer foliage of trees, although its greenish colouration makes it difficult to see well. Greenish above and white below, it has a black head and white throat. With good views you may be able to see a thin white line across the back of the neck, and small patch of bare, bluish skin around the eye.

Where to find The savannas south of Katherine and around Timber Creek are good places to search for this bird, although it is not reliably found in any one location.

☐ Black-chinned Honeyeater

(Golden-backed Honeyeater)
Melithreptus gularis
5½–6½in | 14–17cm

This species is usually found in small parties, and only occurs in the drier savannas of the southern Top End, where it is nomadic, moving around in response to flowering events. Similar to the White-throated Honeyeater but with a black chin and greenish skin around the eye, this bird has a bright golden-green back (hence its alternative common name).

Each of the large honeyeaters shown here is fairly common in the Top End, with a couple of them being particularly common in towns, parks and gardens. The friarbirds, so-called because of their bare-skinned heads, are loud and noisy residents of the Top End's savannas, often moving around in response to flowering events, along with many other honeyeater species.

☐ Little Friarbird *Philemon citreogularis* 10–11½ in | 25–29 cm

Common in wooded habitats across the Top End, this species may be seen in savannas, parks and gardens just about anywhere, often together with Silver-crowned Friarbirds. This is the smallest of the friarbirds and the most nondescript, with bare blue-black skin around the bill and face.

Where to find Easily seen in parks and gardens around Darwin and Katherine.

Little Friarbird

☐ Blue-faced Honeyeater

Entomyzon cyanotis

10–12½ in | 25–32 cm

This species is common in woodland, parks and gardens right across the Top End, where it is usually found in small, active parties, often chasing each other or foraging energetically. Green above and white below, it has a black and white head with a large patch of blue skin around the eye (although this is green in younger birds).

Blue-faced Honeyeater

Where to find Frequently seen in parks and gardens in Darwin and Katherine.

Where to find Common in suburban parks and gardens from Darwin to Katherine, and easily seen in woodlands in Kakadu NP.

Silver-crowned Friarbird

Philemon argenticeps
10–12½in | 25–32cm

A common bird, this species can be found in wooded habitats across the Top End. It is much more widespread than the Helmeted Friarbird, and is smaller, with a silvery mane and crown, and small knob on top of the bill.

Where to find The sandstone subspecies can often be seen at Nourlangie Rock and the coastal subspecies can be seen around Darwin at places like East Point and Buffalo Creek.

Helmeted Friarbird

Philemon buceroides
12½–14in | 32–36cm

There are two distinct populations of this raucous and prehistoric-looking bird in the Top End: one (subspecies *gordoni*) found in mangroves and monsoon forests around the coast; the other (subspecies *ammitophila*) restricted to the sandstone escarpments of Arnhem Land. It has a brown body and bare, black head.

Silver-crowned Friarbird

Helmeted Friarbird

☐ Yellow-throated Miner *Manorina flavigula* 8½–11 in | 22–28 cm

These honeyeaters are very social and extremely territorial. They are usually seen in noisy, active family parties, constantly on the move and giving their high-pitched scolding calls. These parties maintain clear territories, and ferociously exclude other birds from them, with individual miners chasing away smaller birds, or groups of miners aggressively mobbing larger predators. Their territorial defence is so effective that other small birds may be almost completely absent from miner territories. Although widespread and fairly common in woodland across the Top End, these birds are most frequent in the drier woodlands in the south of the region. They are pale grey, with a white rump, dark mask and yellow bill.

Where to find Good places to look include the savannas around Pine Creek and Mataranka.

☐ Weebill

Smicrornis brevirostris
3–3½ in | 7½–9 cm

This species is quite common in savannas and woodlands right across the Top End, particularly away from the coast. It moves around in pairs or small parties, foraging in the outer leaves of eucalypts, and constantly giving its strident call – "*wee-bill, wit-wee-bill*". Australia's smallest bird, this nondescript little species is pale yellow with a small stubby bill.

Where to find Seen regularly in dry woodlands around Katherine and Timber Creek.

☐ Striated Pardalote

Pardalotus striatus
3½–4½ in | 9–11 cm

This unobtrusive little bird is quite common, but can be difficult to see as it spends most of its time foraging in the canopy of eucalypts. A sure sign of its presence is the soft, constant call – "*pit-a-chew, pit-a-chew*". It nests in cavities wherever it can find them, including tree hollows, tunnels in creek banks, holes in bridges and even gaps in brickwork. The plumage is generally pale brownish and yellowish, but the black wings with white flashes, and black cap and prominent white-and-yellow eyebrow are distinctive.

Where to find Possible anywhere there are trees.

☐ Grey-crowned Babbler *Pomatostomus temporalis* 10–11½in | 25–29cm

Common in most open habitats throughout the Top End, these energetic and noisy birds are nearly always found in small family parties. Watching them for a while can be quite amusing, as they seem to get bored easily and are constantly on the move. If they are not foraging actively on the ground or gleaning food from trunks or branches, they will be calling loudly and chasing each other around, seemingly just for fun. Family groups build bulky domed nests of grass and sticks that are used for both nesting and roosting at night, and may have several nests scattered throughout their territory. They breed communally, with all members of the group assisting in building the nest, incubating the eggs and gathering food for the chicks. They are large, long-tailed, brownish birds with a reddish belly, white eyebrow and throat, and longish, downcurved bill.

Where to find Parks and gardens in Darwin, Katherine and other towns throughout the Top End.

Despite their name, woodswallows are not swallows at all, but are most closely related to magpies and butcherbirds.

White-breasted Woodswallow

Artamus leucorhynchus 7 in | 18 cm

An obvious resident in most Top End towns, this bird is most commonly seen in small parties, often sitting huddled up along power lines, or soaring overhead on short, triangular wings. It appears black-and-white, but is actually dark brown above and white below.

Where to find Commonly seen around Darwin and Katherine.

Black-faced Woodswallow

Artamus cinereus 7 in | 18 cm

Common in woodlands throughout the drier parts of the Top End, this bird is usually seen singly or in pairs, sitting on fence posts or power lines, or hawking for insects over the highway. They are grey-brown in colour with an indistinct black mask.

Where to find Often seen beside the highway around Katherine, Mataranka and Timber Creek.

Little Woodswallow

Artamus minor 4½ in | 12 cm

The smallest of the woodswallows, this species is usually seen around cliffs and escarpments, and also the drier savannas of the south and southwest. It has a dark chocolate-brown body and dark bluish-grey wings.

Where to find Look around sandstone escarpments at places like Nourlangie Rock in Kakadu NP and Edith Falls.

White-winged Trille

Lalage tricolor 7 in | 18 cı

This species is common i the dry-season, but mo: birds leave the Top En and migrate to souther Australia over the we season. During the dry season, the majority (birds are dull brown abov and pale below, with som black-and-white edging o the wings. Around Octobe males moult into breedin plumage which is blac above and white below, wit a large white patch on eac shoulde

Where to find
Can be found in woodlands all year around Mataranka and Timber Creek.

White-throated Gerygon

Gerygone olivacea 4½ in | 11 cr

Found across the Top En this species is most commo in the drier savannas in th south of the region, wher the best way to track on down is to listen for it beautiful, lilting 'falling leaf' melody. This tiny bir is green above and yellow below, with a white throa

Where to find Savanna south of Katherine and around Timber Creek.

White-bellied Cuckooshrike

Coracina papuensis

10–11 in | 25–28 cm

Common right across the Top End, this bird's strident "*kiss-ek, kiss-ek*" is a familiar call in woodlands, parks and gardens. It is usually seen in pairs or small groups, often sitting together in the tops of trees. Smaller than Black-faced Cuckooshrike, it is pale grey, with a small black line between the bill and the eye.

Where to find Town parks and gardens from Darwin south to Mataranka.

Black-faced Cuckooshrike

Coracina novaehollandiae

13 in | 33 cm

This well-known species is common right across Australia, including the Top End. It is found in most wooded habitats including savanna woodlands, parks and gardens and is usually seen in pairs or small groups. It is pale bluish-grey with a large black mask, and has a graceful, undulating flight, always giving its wings a little shuffle after landing.

Where to find Easily seen around Darwin and Katherine.

Where to find Woodland around Katherine and Timber Creek.

☐ Varied Sitella *Neositta chrysoptera* 4–5 in | 10–12 cm

These strange birds are always found in small parties, foraging acrobatically along trunks and large branches, often upside-down, investigating cracks and gaps underneath bark for spiders and other invertebrates. They are constantly on the move, spending a few minutes scrutinizing one tree, before all moving on to the next. Although found throughout the Top End, they are most frequent in the drier savannas in the south and southwest. Brown above and white below, the female has an obvious black cap, while the male's is much less distinct.

Varied Sitella

Black-tailed Treecreeper

☐ Black-tailed Treecreeper *Climacteris melanurus* 7–8 in | 18–20 cm

Fairly common in the drier, more open savannas of the central and southern Top End, their loud piping call is usually the first sign that a pair of these birds is around. Treecreepers are Australia's equivalent of woodpeckers, hopping vertically up tree trunks and along branches foraging for invertebrates, before gliding across to the bottom of the next tree and starting again. Dark chocolate-brown, they have a small patch of white streaks on the throat, and a buff wingbar that is obvious in flight.

Where to find Good places to look include the Central Arnhem Road south of Katherine and the escarpment lookout above Timber Creek.

☐ Crested Shrike-tit *Falcunculus frontatus* 6 in | 15 cm

Although common elsewhere in Australia, the subspecies found in the Top End (*whitei*) is one of the region's rarest birds, known from only a handful of locations. It is found in savanna woodlands, with most records from the southern Top End. The bird is usually found using its powerful bill to search for food amongst foliage or loose bark. Its distinctive plumage, green above and pale yellow below, with a black-and-white patterned head, makes it easy to identify.

Where to find Occasionally recorded along the Central Arnhem Road, and Warloch Ponds south of Mataranka.

Rufous Whistler *Pachycephala rufiventris* 6½–7 in | 16–18 cm

The rich and strident song of this beautiful vocalist, particularly the oft-repeated "*eee-CHONG*", is a familiar sound in woodlands right across Australia. It is common and widespread throughout the Top End, including in parks and gardens, and is particularly common in the savanna woodlands of the drier south and southwest where it seems there is always one calling within earshot. An active species, when not singing or defending its territory, it can be found foraging among the outer foliage of trees and shrubs for invertebrates. The males are striking birds, grey above and rufous below with a prominent white throat bordered by a black band. In contrast, the females are a nondescript pale greyish-brown, covered in fine streaks. The females are often confused with other species, but given good views the fine streaking is a useful distinguishing feature.

Where to find In all types of woodland.

☐ Sandstone Shrike-thrush

Colluricincla woodwardi 9–10 in | 23–25 cm

This species has very specific habitat preferences, being restricted to sandstone escarpments in Arnhem Land and also around Victoria River, although it is fairly common where it is found. It is not particularly exciting to look at but is an exquisite songster, its beautiful piping calls echoing from the walls and cliffs of the escarpments. Preferring to sing from an exposed rock, if you hear one calling look carefully for it sitting in the open, although the drab colouring will make it difficult to spot.

Where to find Kakadu NP is the best place to look; they can sometimes be found at Nourlangie Rock, and the escarpment above Gunlom is also a reliable site. Can also be seen from the escarpment walk near Victoria River Crossing.

☐ Grey Shrike-thrush

Colluricincla harmonica 9½ in | 24 cm

Where to find Possible in open woodlands just about anywhere.

Fairly common throughout the Top End, this nondescript species is usually seen foraging low down or on the ground, searching for insects and other invertebrates. It is pale grey with a brownish back, and is best detected by its strident, piping calls.

Where to find Woodlands around Katherine and Mataranka are good places to look.

☐ Olive-backed Oriole

Oriolus sagittatus 11 in | 28 cm

Although present year-round, the population of this fairly large bird increases during the dry-season with an influx of migrants from southern Australia. It is quite common in savannas across the Top End, particularly in the drier southern areas of the region. Both sexes are olive green above, have a pale breast covered in prominent olive-green streaks, and an orange bill.

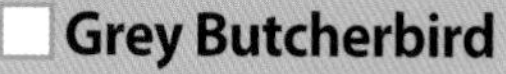

Grey Butcherbird

Cracticus torquatus
(Silver-backed Butcherbird)
9½–11 in | 24–28 cm

Where to find Good places to search include the cemetery at Adelaide River and Umbrawarra Gorge near Pine Creek.

The smallest butcherbird and probably the least common, this species is found in woodlands, parks and gardens throughout the Top End, often close to water. The white throat is a good feature for separating it from the similar Pied Butcherbird.

Pied Butcherbird

Grey Butcherbird

Pied Butcherbird

Cracticus nigrogularis
12½–13½ in | 32–34 cm

Common right across the Top End, these familiar black-and-white birds are a frequent sight on power lines and telephone poles throughout the region, and their beautiful, piping call is a regular component of the morning chorus. Like other butcherbirds they eat almost anything they can catch with their powerful, hooked bills, and are so called because of their habit of wedging their prey in a fork between branches, or impaling it on a stick before dismembering it.

Where to find Easily seen, and one of the most obvious birds in the Top End.

Australasian Magpie

(Australian Magpie)
Gymnorhina tibicen
14–17½ in | 36–44 cm

The familiar large black-and-white bird in southern and eastern Australia, this species is much rarer in the Top End. It is only found regularly in the drier areas towards the south of the region, and is usually seen walking around and foraging on the ground in parks and on playing fields and road verges.

Where to find Look for this species south of Katherine, particularly around Mataranka.

Where to find Easily seen almost anywhere across the Top End.

☐ Willie-wagtail *Rhipidura leucophrys* 7½–8½ in | 19–22 cm

These lovely black-and-white fantails are one of the favourite birds of many Australians. Common and familiar residents in most open habitats across the Top End, they have adapted particularly well to human settlement and are common around cities and towns. Continuously active, they are usually seen darting about to catch insects and repeatedly wagging their long black tails back and forth. Willie-wagtails are very vocal birds, calling throughout the day, and occasionally during the night. Their ratchetty call is a well-known sound, and often quoted as "*sweet-pretty-creature*". Surprisingly for their small size, these birds are fearless, with pairs vigorously defending their territories, swooping and scolding much larger birds, dogs, cats and even people! Willie-wagtails build a beautiful, small, cup-shaped nest that is placed on a horizontal branch and bound tightly with spiderwebs. Enterprising pairs will often build their nest on the rafters or pipes under the eaves of a house. They usually raise around three or four young, and the tiny nest can become very crowded if there are four nestlings about to fledge. Easily recognised, they have a black head and wings, a long black tail, a white breast and a small white eyebrow.

Where to find Anywhere there is open space.

☐ Magpie-lark *Grallina cyanoleuca* 10½ in | 27 cm

A familiar species across the Top End, these striking black-and-white birds are usually seen in pairs, foraging on the ground in parks, on playing fields or beside roads. They build a cup-shaped nest out of mud, which they line with grass and other soft material, sometimes placing it on rafters or pylons around buildings. The sexes can be told apart, males having a white eyebrow and white cheek patch, whereas females have a vertical black line through the eye.

Paperbark Flycatcher

Myiagra nana 6½in | 17cm

Common in most wooded habitats throughout the Top End, this bird's strident two-note whistle, "*choo-wee, choo-wee*" or strange grinding call, are usually the first clues to its presence. Both sexes are glossy blue-black above and white below.

Where to find Copperfield Dam near Pine Creek is a good place to look for this species, but it can also be seen throughout Kakadu NP and around Katherine.

Leaden Flycatcher

Myiagra rubecula
6–7in | 15–17cm

This little flycatcher is common across the Top End in most wooded habitats and has the curious habit of constantly flicking its tail up and down. The female is very similar to Broad-billed Flycatcher (*page 117*), being grey above and white below with an orange throat; however, the male is quite different, being bluish-grey above and white below with a blue-grey hood. Broad-billed Flycatchers are always found near water, but as both species usually occur in pairs, finding a male is the best way of being sure of your identification!

Where to find Quite common in woodlands throughout Kakadu NP.

Lemon-bellied Flycatcher

(Lemon-bellied Flyrobin)
Microeca flavigaster
5½ in | 14 cm

This small flycatcher is usually seen sitting high in the canopy, where it sallies for insects from a prominent branch. It also uses these high perches to broadcast its pretty, bubbling song. This species is common in most habitats across the Top End, and is often found close to water. It is a rather dull bird, olive-green above and pale yellow below, and has a white throat.

Where to find Look in woodland around the margins of billabongs in Kakadu NP.

Jacky Winter

Microeca fascinans
5½ in | 14 cm

This bird is fairly common in the drier areas of the Top End, and is usually seen sitting on a low snag or stump, sallying out or dropping onto insects on the ground. It is a fairly nondescript robin, with the best identification feature being its unusual habit of constantly wagging its white-edged tail from side to side while perched.

Where to find Search in the drier savannas around Mataranka and Timber Creek.

Where to find There are usually a group or two around Mataranka.

☐ Apostlebird *Struthidea cinerea*

11½–12½ in | 29–32 cm

As the name suggests, these scruffy grey birds are usually found in small groups, wandering around on the ground and pumping their tails up and down, or sitting together in a tree. They are noisy, constantly giving harsh grating calls, or a loud, descending whistle. All members of the group help build a mud nest and assist with raising the young. Apostlebirds are only found in the southern and southwestern Top End, where they are often seen along roads and around towns.

☐ Torresian Crow *Corvus orru*

19–21 in | 48–53 cm

The only crow found in the Top End, these birds are common and obvious residents around towns and cities. They have adapted well to human habitation and can often be seen scavenging for scraps or handouts around shopping centres, parks and gardens.

Where to find Easily seen in suburban areas of Darwin and Katherine.

☐ Australasian Bushlark

(Australasian Lark; Singing Bushlark)
Mirafra javanica
5–6 in | 13–15 cm

Quite common across the Top End, this bird is found in open fields, airfields, grasslands and floodplains. It is best seen in the breeding season, when males give fluttering display flights while constantly singing, high above their open habitat. The similar **Australasian Pipit** *Anthus novaeseelandiae* (not illustrated) also occurs in the Top End, but is longer and more slender, with a more graceful, undulating flight.

Where to find Good places to look include the margins of Knuckey Lagoons in Darwin, and Timber Creek airfield.

☐ Golden-headed Cisticola

Cisticola exilis
4–4½ in | 9–11 cm

Common across the Top End, this tiny bird can be found in almost any wetland or tall grassy area, where the distinctive call is often heard – a strange buzz followed by a loud, clear note, something like "*bzzzzt-PLINK!*" It is pale below and has a streaky back and golden-brown head, fine dark streaks developing on the crown outside the breeding season.

Non-breeding

Breeding

Where to find These birds are easy to see at Holmes Jungle Swamp and Knuckey Lagoons in Darwin.

Mistletoebird

Dicaeum hirundinaceum

4–4½ in | 10–11 cm

Although found in nearly all habitats across the Top End, this bird unfortunately spends much of its time in the canopy, making the beautiful red-breasted males difficult to see. Females are much plainer than males and have a small patch of orange under the tail. However, once the high-pitched "*seew*" call of this bird is learned, you will be surprised how common it is.

Where to find Possible in woodlands anywhere, but particularly near water.

Where to find Look for this species around Katherine in the dry-season.

☐ Fairy Martin *Petrochelidon ariel*

4½ in | 12 cm

Not always easy to find, the Fairy Martin is a seasonal visitor that is most common during the dry-season, primarily in the drier areas of the southern Top End. It is often seen near water and shows a particular fondness for culverts or bridges over small creeks. Very similar to the Tree Martin, this species has a whiter rump and a pale chestnut crown.

Where to find Can be seen anywhere there are trees, including in urban areas.

☐ Tree Martin *Petrochelidon nigricans*

5½ in | 14 cm

Rarely seen perched, this graceful master of the air is usually seen in loose flocks or groups hawking for insects over open ground. It is the most common 'swallow' of the region, and is particularly numerous during the dry-season when birds from southern Australia migrate to the Top End. It is dark above and pale below, with a short, square tail.

Fairy Martin

Tree Martin

☐ Double-barred Finch

Taeniopygia bichenovii

4–4½ in | 10–11 cm

This distinctive finch with its owl-like face is usually found in pairs or small parties, feeding on the ground or flying swiftly to cover. They are one of the most common finches in the Top End, and are regular visitors to small waterholes.

Where to find Easily seen around Darwin, Katherine and Timber Creek.

Where to find Holmes Jungle Swamp is a good place to see this bird close to Darwin, but they can be found in suitable habitat throughout the region, right down to Mataranka.

☐ Crimson Finch *Neochmia phaeton*

5–5½ in | 12–14 cm

These beautiful little blood-red finches are quite easy to see in the Top End. Usually found in small family groups, they spend their time foraging on the ground or flitting around in dense vegetation, almost always close to water. They have a fondness for *Pandanus* trees, a type of small palm often growing near water, and they sometimes build their bulky, domed nests among the spiky fronds.

Where to find In the late dry-season Timber Creek is a good place to search, particularly in the riverside grasses at Policeman's Point, or around the airfield. They can also be found in the long grass at the Old Victoria River Crossing.

☐ Star Finch *Neochmia ruficauda* 4½ in | 11 cm

One of the more difficult finches to find in the Top End, Star Finches are nomadic, moving around in response to local conditions. Found mainly in the southwest of the region, in the late dry-season they gather in fairly large flocks and are usually seen in areas of long grass, often close to water. They are greenish-brown above and pale yellow below with a red face. The breast is peppered with small white spots, hence their name.

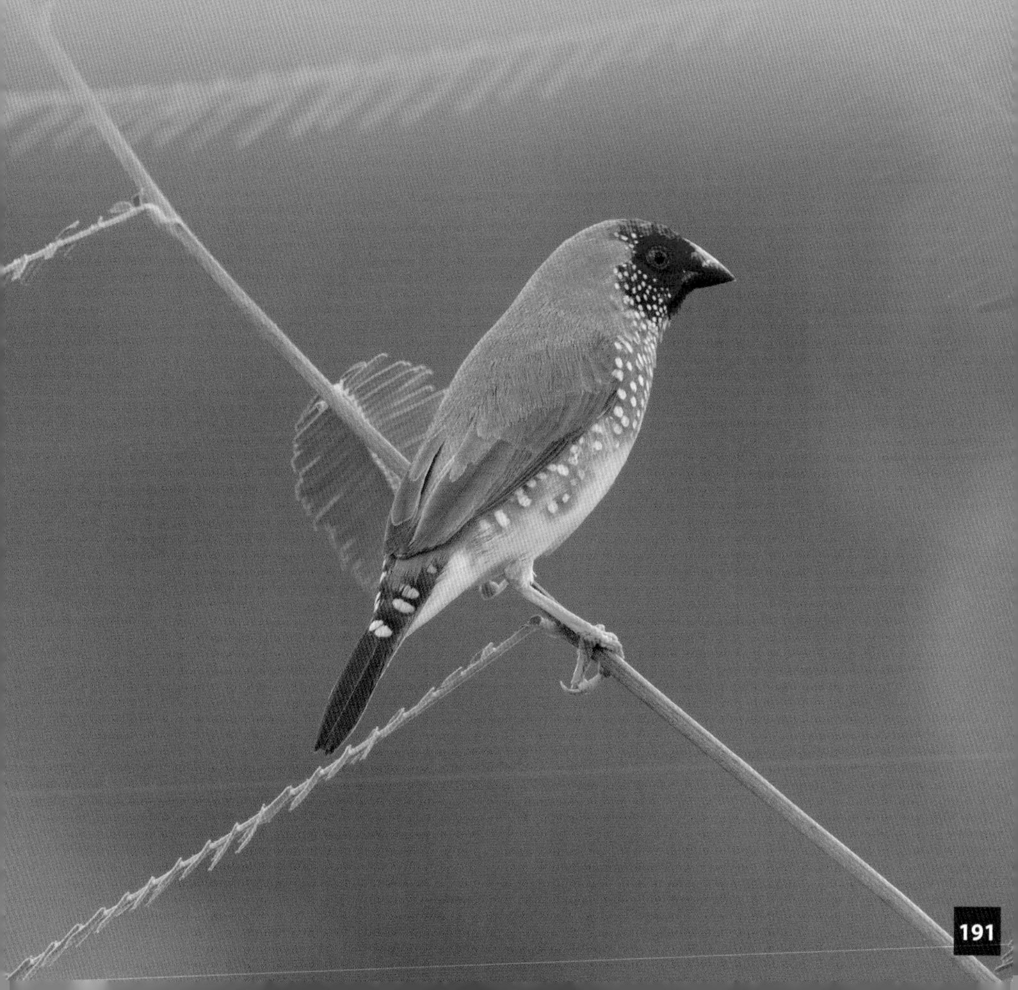

Red-faced
Gouldian Finch
Black-faced
Long-tailed Finch
Masked Finch

☐ Gouldian Finch *Erythrura gouldiae* 5–5½in | 13–14cm

Ask any birdwatcher visiting the Top End for the first time which species they most want to see, and there is an excellent chance that this beautiful bird will be at the top of their list. Seen most often in the grassy woodlands of the southern Top End, Gouldian Finches are usually found in small flocks, often of mixed adults and juveniles, which are much paler and less colourful than the adults. They often associate with other finch species, and it is not unusual to find four or more species of finch drinking together at waterholes late in the dry-season. Once common, these birds underwent a severe population decline, and although the population is still quite small, it seems to be slowly recovering. The reason for the decline is unknown, but may be related to changing fire regimes. The Gouldian Finch has three colour 'morphs': most birds have black faces, but a few have red faces and every now and then you may see one with a yellow face.

Where to find The Katherine region is the best place to search for this species, particularly along the road into Edith Falls. The southwestern Top End is another good place to look, including the airfield at Timber Creek and along the first few kilometres of the Buchanan Highway.

☐ Long-tailed Finch *Poephila acuticauda* 6–6½in | 15–16cm

Fairly common in grassy savannas across the Top End, including around Darwin, this beautiful species is often found in company with Masked Finches, and both species are regular visitors to small waterholes at the end of the dry-season. Throughout most of the Top End, Long-tailed Finches have red bills, but in the far southwest of the region you may occasionally see birds with yellow bills. Although these individuals may look similar to Masked Finch, that species is much browner, and never has the pale grey head and black throat of Long-tailed Finch.

Where to find Woodlands around Katherine and Timber Creek.

☐ Masked Finch *Poephila personata* 4½–5½in | 12–14cm

This species is usually found in pairs or small parties, and sometimes quite large flocks in the late dry-season. It is often seen in mixed flocks with Long-tailed Finches, and will visit waterholes at dawn and dusk to drink. Fairly common in grassy savannas across the Top End, like other finch species it will move around as it searches for the best conditions. As a result it can be numerous in some years but quite scarce in others. An attractive little finch, it is brown with a large yellow bill and small black mask.

Where to find Seen most regularly in the south and southwest of the region, particularly around Katherine and Timber Creek.

☐ **Chestnut-breasted Mannikin** *Lonchura castaneothorax* 4½–5 in | 11–12 cm
(Chestnut-breasted Munia)

These attractive little birds can be found across the Top End, and are often seen in quite large flocks numbering hundreds of birds. They are most often seen in areas with long grass, and can be quite common along river margins or around the edges of wetlands. A stunning little bird, adults have brown wings and a white belly, a black mask and chestnut breast-band, bordered below by a bold black line. Flocks are usually made up of a number of adults and quite a few young birds, which are plainer brown with a greyish head, and lack the striking breast pattern. Also keep an eye out for Yellow-rumped Mannikins which often occur in mixed flocks with this species.

Where to find Good places to search include Holmes Jungle Swamp in Darwin, and around Timber Creek at sites like Policeman's Point and the airfield.

☐ **Yellow-rumped Mannikin** *Lonchura flaviprymna* 4½–5 in | 11–12 cm
(Yellow-rumped Munia)

These birds are nearly always found in mixed flocks with Chestnut-breasted Mannikins, but are much less common. They often occur in only ones and twos, so you must carefully check through flocks of mannikins if you are to find them. Occasionally seen as far north as Darwin, they are usually found only in the southwestern Top End, between Victoria River Crossing and Kununurra. Like Chestnut-breasted Mannikins, they prefer areas of long grass, and are most regularly found along the margins of rivers including the Victoria and Ord Rivers. Brown above and yellowish below, they have a white hood, yellow rump and tail, and stout, bluish bill.

Where to find Focus your search around Timber Creek, particularly at the airfield, and also in long grass along the river at Policeman's Point.

☐ **Pictorella Mannikin** *Heteromunia pectoralis* 4½ in | 11 cm
(Pictorella Munia)

The most difficult finch in the Top End to track down, this species prefers drier habitats and may turn up almost anywhere in the south of the region, but is only regularly seen in the dry savannas around Timber Creek. As with most finches, the best strategy for seeing it is to find a small waterhole at the end of the dry-season and wait for birds to come in to drink. Although not a gaudy species, it is nonetheless very attractive, adults being brown above with a black mask and having a black-and-white patch on the upper breast. Young birds are quite nondescript, being just pale brown, and are easily confused with the young of other finch species, many of which look quite similar. A good feature to look for is the bill, which is much larger than in the other species of finch.

Where to find Small waterholes along the Buchanan Highway and Bullita Access Roads near Timber Creek are good places to look.

Chestnut-breasted Mannikin

Yellow-rumped Mannikin

Pictorella Mannikin

Mammals

Mammals are well represented in the Top End and the region is home to a wide variety of incredible species. Some are obvious, with species like the Agile Wallaby and Black Flying-fox being widespread and easily found, but many are nocturnal and secretive, and only seen by the keen observer willing to get out searching. This section doe not provide an exhaustive list of all the mammals that occur in the Top End, but covers those species that you are most likely to encounter, and a few that require a bit of effort and good luck to find.

Agile Walla

Black-footed Tree Rat

Black Flying-fox

☐ **Short-beaked Echidna** *Tachyglossus aculeatus* 12½–17½ in | 32–44 cm
(Spiny Anteater)

Usually just called 'Echidna' (although there are three species of long-beaked echidna in New Guinea), in evolutionary terms this species is among the most ancient mammals in the world. Along with the Platypus *Ornithorhynchus anatinus* of eastern Australia, Echidnas are the only mammals that lay eggs. The spherical eggs are only small, about 13–15 mm in diameter and have leathery shells – they are not brittle-shelled like a chicken's egg. The female keeps her egg in a small rear-facing 'pouch', really just a small fold of skin, until it hatches. The baby Echidna is called a 'puggle', and stays with the mother until its spines start to grow. Once this happens the female leaves her youngster behind in a burrow, under a log or in some other shelter while she forages, returning every few days to suckle it. Adult Echidnas feed exclusively on ants and termites, tearing apart the nests with their powerful feet and claws before using their long, sticky tongues

good memory that enables them to recall the locations of ants' nests and termite mounds across their large territories. They are wary creatures, and if startled on open ground will immediately curl themselves up into a tight ball, spines on the outside. Sometimes, if they are on soft or sandy ground, they will dig vertically down into the soil with surprising speed, leaving just their spines exposed, and not coming out until they are sure danger has will see it raise its long snout first to smell for danger, before coming back up and waddling off. Echidnas are most active during the dry-season, particularly during the cooler periods of the day and at night.

Where to find Less common in the Top End than other parts of Australia but may occur almost anywhere, although rarely seen.

The size of a small cat, these active omnivorous marsupials are nocturnal, emerging after nightfall to forage both on the ground and in trees. They feed on insects and other invertebrates, small mammals, birds, frogs, reptiles and even fruit – whatever is most abundant at the time. Northern Quolls are very short-lived, primarily due to their unusual breeding ecology. All females in the population come into oestrus (ready for mating) within a two-week period early in the dry-season, and the males spend this time doing nothing except trying to mate with as many females as possible. The males do not eat during this time and expend so much energy competing with other males that they quite literally starve themselves to death – all males dying within a couple of weeks of mating, at only one year old. Within a few weeks of mating the females give birth to a litter of 7–8 young, which attach to her teats and stay with her for a few months until the young get so large that she has to leave them in a den while she forages. After about five months they leave the den, eventually dispersing, with most breeding females dying soon after, at less than two years old. Only about 30% of females survive to raise young for a second year. Like so many of Australia's small marsupials, the Northern Quoll has suffered dramatic population declines, with the primary cause in the Top End being the arrival of the introduced and poisonous Cane Toad (*page 263*).

Where to find Once found throughout the Top End but now restricted to savannas and rocky escarpments in the west of the region and very difficult to find; still occasionally seen in Kakadu NP, and a sizeable population remains on Groote Eylandt.

Northern Brush-tailed Phascogale *Phascogale pirata* 13–16 in | 33–41 cm

This small, carnivorous marsupial is nocturnal, spending most of its time in the canopy, dashing about on tree trunks and branches searching under bark and in hollows for insects, spiders and small vertebrates, although it will also feed on nectar in flowering eucalypts. Like many small species in the family Dasyuridae (including the Northern Quoll), their breeding biology means they are very short-lived, with males dying off after the breeding season, and most females dying after they have finished raising their first litter of young. Phascogales are very shy animals, and if you are lucky enough to find one, you will probably only get a very brief view. They avoid spotlights, running to the back side of the tree and staying out of sight. Like the Northern Quoll, this species has also declined severely in the Northern Territory. The reasons for this decline are unknown, but could be linked to the arrival of Cane Toads and/or the increasing populations of feral cats.

Where to find Historically very common but now rarely recorded except in some parts of Kakadu NP where spotlighting at night may provide a chance of seeing one.

Northern Brown Bandicoot *Isoodon macrourus* 19–22½ in | 48–57 cm

About the size of a rabbit and pale brown with a long snout and short tail, these marsupials spend the day sheltering in a shallow nest covered with sticks, grass and leaves. They emerge at night to forage for insects, earthworms, fruit and fungi, using their acute sense of smell to sniff out food. They often dig for food underground, leaving small conical holes – a sure sign that they are around. These animals have incredible rates of reproduction: females are able to breed at only 3–4 months old, and have one of the shortest gestation periods of any mammal at only 12–13 days! The 2–4 babies are only about a centimetre long when born, naked and blind, but after attaching to a teat in the mother's pouch they grow quickly on the very rich milk. After two months the young leave the pouch, by which time the female may already be pregnant again. In this way they are able to breed up to three times a season, and it is therefore not surprising that these creatures are quite common across the Top End, even in suburban areas.

Where to find Look for this species at night in suburban parks and gardens or around campsites, particularly near areas with long or dense grass.

Common Brushtail Possum *Trichosurus vulpecula* 23½–30½in | 60–77cm

About the size of a small cat, these possums are greyish above and whitish below with a chunky head and large ears. The tail is almost as long as the body and, despite the name, not noticeably brushy in animals from the Top End. They are patchily distributed in woodlands across the region but seem to do well in urban environments. They spend most of the day curled up in a tree hollow, roof of a house or other dark recess but emerge at night to climb around in tree canopies where they feed on leaves, flowers and fruit. Young possums spend about five months in the pouch and then ride around on their mother's back before they grow too large. Found from the Top End south to Tasmania, this species is a good demonstration of the consistent differences that exist between populations from northern or tropical climates, and southern, more temperate climates. Common Brushtail Possums from the Top End are pale grey, have sparse, short fur, and weigh about 1·5 kg. By comparison, brushtails from Tasmania are almost black, have thick, woolly fur, and are large, weighing up to 4 kg (all adaptations to the cooler climate).

Where to find Quite common in suburban Darwin, where it can be found in parks like East Point. Also search for it around campgrounds in Kakadu NP.

This unusual species is restricted to sandstone escarpments or rocky outcrops around the central Arnhem Land escarpments and also the far southwestern Top End. Unlike most large possums, Rock Ringtails live in small groups of up to ten animals, with a very strong social bond. During the day they rest together in small caves, deep rock crevices and sometimes on protected edges, emerging after dark to troop across the rocky outcrops to feed in trees, where they eat mainly leaves, but also fruit and flowers when available. Both males and females play an important role in raising the young and protecting the group. Males in particular will aggressively defend the group, warning other members of danger both vocally and by slapping their tail on the ground. The young are very well cared for and protected by the parents who have even been seen forming a bridge across cracks in the rock or between trees so the young can run across their backs. These possums are patchily distributed across the Top End, seemingly absent from suitable habitat but common in other areas. Often the best sign of their presence is their distinctive scats (faeces), which are about 15–25 mm long and shaped like a bent cigar.

Where to find Found around some of the sandstone escarpments in Kakadu NP including Bardedjilidji, Nourlangie and Gunlom, but be aware that access to some of these areas is not permitted at night.

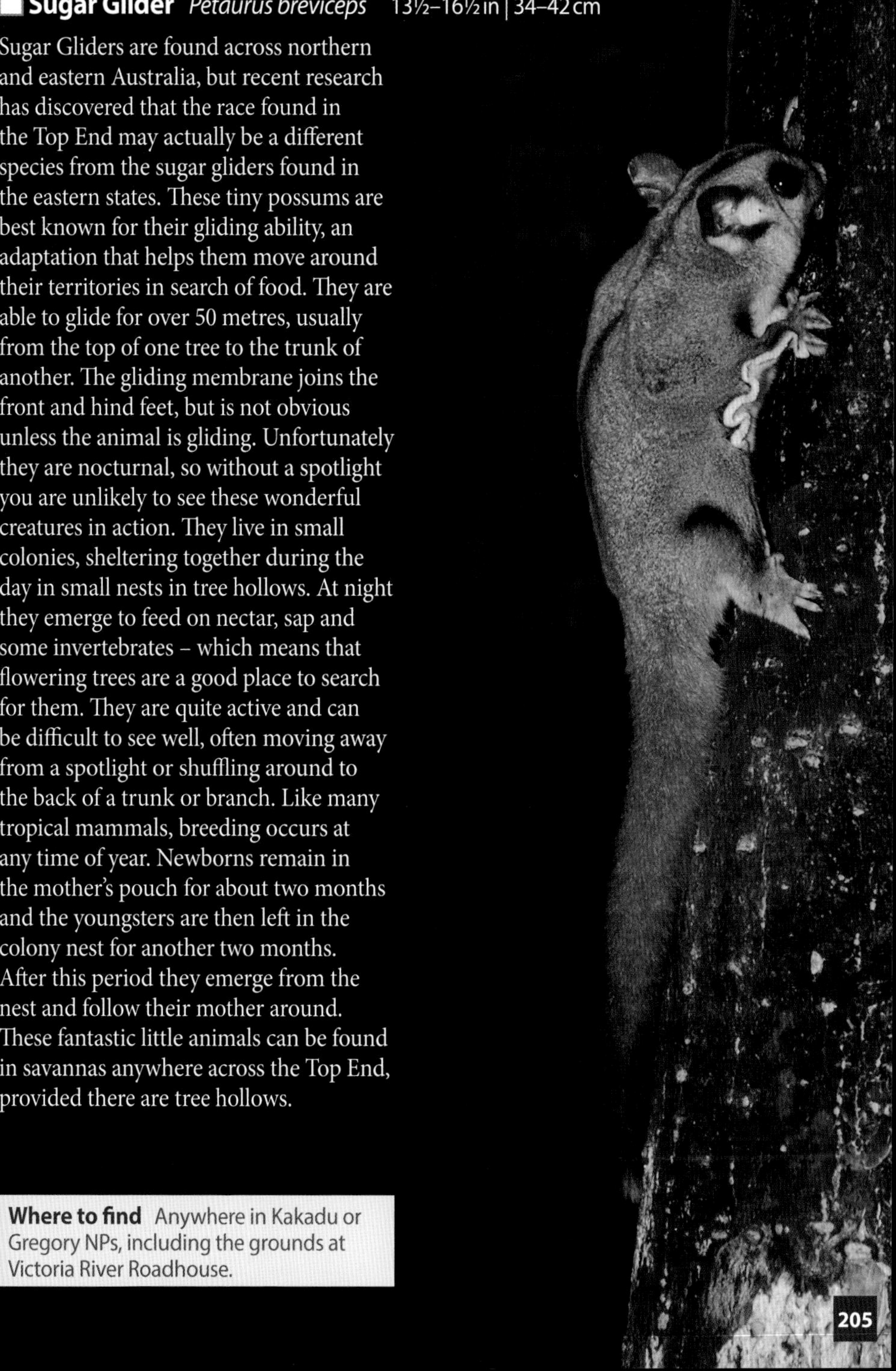

Sugar Glider *Petaurus breviceps* 13½–16½ in | 34–42 cm

Sugar Gliders are found across northern and eastern Australia, but recent research has discovered that the race found in the Top End may actually be a different species from the sugar gliders found in the eastern states. These tiny possums are best known for their gliding ability, an adaptation that helps them move around their territories in search of food. They are able to glide for over 50 metres, usually from the top of one tree to the trunk of another. The gliding membrane joins the front and hind feet, but is not obvious unless the animal is gliding. Unfortunately they are nocturnal, so without a spotlight you are unlikely to see these wonderful creatures in action. They live in small colonies, sheltering together during the day in small nests in tree hollows. At night they emerge to feed on nectar, sap and some invertebrates – which means that flowering trees are a good place to search for them. They are quite active and can be difficult to see well, often moving away from a spotlight or shuffling around to the back of a trunk or branch. Like many tropical mammals, breeding occurs at any time of year. Newborns remain in the mother's pouch for about two months and the youngsters are then left in the colony nest for another two months. After this period they emerge from the nest and follow their mother around. These fantastic little animals can be found in savannas anywhere across the Top End, provided there are tree hollows.

Where to find Anywhere in Kakadu or Gregory NPs, including the grounds at Victoria River Roadhouse.

A frequently asked question is "What is the difference between a kangaroo and a wallaby?" Really, the difference is arbitrary as most belong to the same scientific genus (*Macropus*). 'Kangaroo' is the term given to the larger macropods (the term used for a member of the marsupial family Macropodidae), while 'wallaby' is the name given to the smaller species. Wallaroos are somewhere in the middle; large, stocky species, not as lanky as kangaroos, but larger than wallabies

■ Agile Wallaby *Macropus agilis* 51–63 in | 1·3–1·6 m

The common small macropod across the Top End, these little wallabies are quite abundant in savanna and grasslands across the Top End, especially in the south and southwest. They are quite shy, often hopping off very quickly when disturbed with a distinctive gait: the front part of the body is held upright, the stiff 'U'-shaped tail is held low and horizontal, and the short arms are held straight out in front. Like most macropods they are mainly nocturnal, spending the day resting in long grass or scrub and emerging at dusk to feed in open areas, although on overcast days they will forage during the day. Agile Wallabies are essentially solitary, although animals will gather in loose groups at remaining food sources when food is scarce. They are grazers and their preferred food is grasses and herbs, which are plentiful during the wet-season but as the dry-season progresses and less grass is available, they turn to eating more leaves, fruits, flowers and roots.

Where to find In Darwin there is a healthy population at East Point. Victoria River Crossing is an excellent place to see them, and they are also common at Policeman's Point near Timber Creek.

Northern Nailtail Wallaby *Onychogalea unguifera* 43½–55 in | 1·1–1·4m

This small-bodied wallaby has a very long tail, greater than the length of its body, with a small horny tip that can only be seen from very close range. Historically there were three species of 'nailtail' wallabies found in Australia, but one is now extinct and the other, the **Bridled Nailtail Wallaby** *Onychogalea fraenata*, was feared extinct before a tiny population of a few hundred animals was rediscovered in Queensland in 1973. Only the Northern Nailtail Wallaby remains widespread, perhaps because the introduced **Red Fox** *Vulpes vulpes* (not illustrated), an efficient predator of small wallabies, is not prevalent in northern Australia. It is generally uncommon in savanna and grasslands across the Top End, but is moderately common in the drier savannas south of Katherine. It is a solitary animal, spending the day sheltering in dense cover before emerging at night to feed mainly on herbs, with some grasses and fruit. Very similar to the Agile Wallaby, it can be difficult to identify without experience – though the Northern Nailtail Wallaby hops quite differently, holding its body low and horizontal, with the much longer tail trailing behind.

Where to find Difficult to find, the best chance is spotlighting along roads in the south or southwest of the region, particularly south of Mataranka or around Top Springs, Victoria River and Timber Creek.

☐ Antilopine Wallaroo *Macropus antilopinus* 55–82½ in | 1·4–2·1 m

All three species of wallaroo are found in the Top End, with Antilopine Wallaroo being the most common. It is a gregarious species, usually found in small groups resting in the shade or emerging to graze late in the afternoon. Unlike many tropical mammals, breeding is seasonal, with most joeys (babies) being born near the end of the wet-season before spending the dry-season in the pouch and emerging at the start of the next wet-season when grass is plentiful. Although not as common as the Agile Wallaby (*page 206*), Antilopine Wallaroos are found in many of the same areas. They are much larger and hop quite differently, with a relaxed, upright gait, rather than the frantic horizontal 'bolting' of wallabies. Antilopine Wallaroos have a distinctive mule or horse-like head, with a large black nose. Males are much larger than the greyish females, and are reddish above and whitish below.

Where to find Look for this species along road edges throughout Kakadu NP, and in the savannas of the southern and southwestern Top End.

☐ **Common Wallaroo** *Macropus robustus* 55–79 in | 1·4–2·0 m
(Euro)

Found right across Australia, this large, stocky-bodied macropod is fairly common in suitable habitat in the Top End. It is variable in colour but generally reddish or reddish-grey in the Top End. Males are much larger than females and have large, powerful forearms. A solitary animal, it is usually found in steep, rocky areas, where it hops around carefully with a distinctive hunch-backed posture. Common Wallaroos share a curious breeding ecology with several other macropods. When born, they are only the size of a jellybean, completely naked and blind with no hind limbs and tiny forearms. They crawl through the mother's fur to the pouch and attach themselves to a teat. At this point the female enters oestrus again, but when the new egg is fertilized it enters 'embryonic diapause', remaining viable but not developing. The suckling young continues developing in the pouch, but when it becomes too big, vacates the pouch and remains close to its mother. At this point the 'paused' embryo begins developing again, being born soon after and making its way to the pouch. Another embryo is then fertilized, and in this way the female may have three offspring at one time: a 'paused' embryo, one newborn in the pouch and one out of the pouch. If at any point the joey in the pouch dies, the female can release the embryo from diapause and replace it, a fascinating adaptation to the harsh climate of Australia.

Where to find Search along steep, rocky hillsides; the escarpment above Timber Creek is a good place to look.

☐ Black Wallaroo *Macropus bernardus* 47–55 in | 1·2–1·4 m

One of only a handful of mammals found only in the Top End, this species is restricted to the sandstone escarpments of western Arnhem Land. Females are grey-brown, while the larger males are entirely dark chocolate-brown or black. They are actually quite common in their rocky habitat, but are very wary. During the day they spend most of their time sheltering under ledges or large trees, emerging at night to feed on grass, leaves and fruit. They are usually seen when startled from their resting place, generally allowing only fleeting glimpses as they bound up steep escarpments or among large boulders before disappearing. A good strategy for seeing these animals is to find a secluded location in the early morning or late afternoon from where you can see a large area of escarpment, and wait quietly and look for an individual on the move.

Where to find Search for this species in Kakadu NP. The Gunwarddehwarde Lookout at Nourlangie Rock is a good place to find them, but perhaps the best location is the escarpment at the top of Gunlom waterfall.

Rock-wallabies 29½–43½ in | (75–110 cm)

These small wallabies are the northern representatives of a complex of small rock-dwelling wallabies found throughout Australia. Until late 2014, they were considered the same species, Short-eared Rock-wallaby, but scientists recently determined that animals from the northern and eastern Top End (including Litchfield and Kakadu NPs) were in fact a different species from those in the Victoria River region of the southwestern Top End, renaming them 'Wilkin's Rock-wallaby'. The similar but much smaller **Nabarlek** *Petrogale concinna* (not illustrated) is also found in the Top End, although it is much rarer. Both rock-wallabies are restricted to rocky outcrops, cliffs and escarpments, though are quite common in suitable habitat. They are incredibly agile, bounding rapidly with surprising ease even across vertical rock faces. During the day they rest in dens, which may be deep within rock piles or caves, emerging at dusk to feed on grasses, fruits and leaves. Very wary, they rarely stay in the open for long, usually hopping away to another hideout when disturbed. The best times to find either species are early morning (when they may be seen sunning on ledges) and late afternoon (when they emerge to feed).

☐ Wilkin's Rock-wallaby
Petrogale wilkinsi

Found in the northern and eastern Top End, this species is variable in colouration and pattern, being brownish, often with a small black crescent behind the forearms, and a long tail that has a brushy black tip.

Where to find Often seen at Litchfield NP and Bardedjilidji in Kakadu NP.

☐ Short-eared Rock-wallaby
Petrogale brachyotis

Found in the southwestern Top End, this small wallaby is much plainer than Wilkin's Rock-wallaby.

Where to find Common on the Victoria River Roadhouse escarpment walk.

Short-eared Rock-wallaby

Wilkin's Rock-wallaby

☐ Black Flying-fox *Pteropus alecto* 10–11 in | 25–28 cm

Australia's largest bat, it is hard to miss these raucous creatures if they are around. They roost together during the day in noisy colonies that may number tens of thousands, usually in thick vegetation along rivers or creeks. These colonies are hives of constant activity, with continually screeching and squabbling bats providing endless entertainment. At night they stream out across the landscape in lumbering flight, searching for fruiting or flowering trees. Individual bats may cover more than 20 km in a single night. A heavily fruiting tree may attract many bats, which crash-land into the canopy before clambering around searching for fruit and continuing to fight with one another. Mothers give birth during the dry-season and initially the young bat clings to the mother's belly, even as she flies about at night foraging. After it becomes too large the youngster stays behind in the colony until eventually it is able to fly about on its own. These bats do particularly well in suburban areas, where the fruiting and flowering trees that have been planted provide a great food source.

Where to find There are large roosting colonies at Howard Springs close to Darwin, at Katherine Gorge and at Timber Creek, but they can be seen in most places at night.

Little Red Flying-fox *Pteropus scapulatus* 8–9½ in | 20–24 cm

This small flying-fox has a reddish-brown body, often with a paler neck and darker head and translucent, reddish wings. Like Black Flying-foxes, this species sometimes forms very large colonies, up to several hundred thousand animals, but may also be seen in small 'camps' (groups) of just a few bats roosting in tight clusters close to water. Much less predictable than the larger Black Flying-fox, the Little Red Flying-fox moves around the landscape more randomly. This is because nectar is its main source of food and different parts of the forest flower at different times. Large numbers of bats can therefore arrive and occupy an area for a short period before moving on. Its reliance on nectar means that this species plays an important role in the ecology of northern Australia's savannas, as it is one of the chief pollinators in these woodlands.

Where to find Can be found anywhere across the Top End (though tends to move south in the dry-season), and can often be found by spotlighting around flowering trees after dark.

Microbats

Yellow-bellied Sheath-tailed Bat

'Microbat' is a term given to the small insectivorous bats that emerge at dusk and are often seen fluttering around in the fading light of day. Apart from the large flying-foxes, all bats in the Top End belong to this group of bats. Many of them are impossible to identify unless you can trap them and examine them 'in the hand', but some can still be identified if you know what to look for. Most species spend their days roosting in dark places, such as caves, abandoned mine shafts, tree hollows, under bark or even in houses, before emerging at dusk to hunt. Each has a different hunting strategy: some of the larger, long-winged microbats fly about high and fast above the canopy searching for insects; some species forage over open areas such as creeks or clearings; while other smaller, more manoeuvrable species forage below the canopy, zig-zagging in and around branches and trunks.

Microbats are best known for their ability to echolocate – they emit a series of high-pitched pulses and can detect these pulses when they rebound off whatever objects are nearby, enabling them to navigate in complete darkness. The frequencies of the pulses emitted by most species are too high for humans to detect, but they can be recorded using special electronic equipment and analyzed, with each species having a very distinctive call which, much like bird calls, are unique. The calls of a handful of species are audible to the human ear, such as the metallic "*tik...tik...tik...tik*" of the **Yellow-bellied Sheath-tailed Bat** *Saccolaimus flaviventris*, or the loud "*chirp*" of the **Ghost Bat** *Macroderma gigas*.

Their ability to echolocate, and the saying 'blind as a bat', has promoted the common misconception that bats are blind. All species of bat can see and some species are able to see quite well. While they use echolocation to move around in the dark, much of their actual hunting is done visually. Most microbats feed on insects, although the **Ghost Bat** is a carnivore that eats other vertebrates including lizards, frogs, birds and other bats, catching them before returning to a feeding perch where it dismembers and eats its prey.

Identifying microbats is usually quite difficult without being able to trap and examine them in detail, although some species can be identified if you can keep them in a spotlight long enough. **Ghost Bats** are Australia's largest microbat and have a pale body, very long ears and large eyes. The **Yellow-bellied Sheath-tailed Bat** is also relatively large, is white below and black above and has very long, narrow wings. The smaller **Orange Leaf-nosed Bat** *Rhinonicteris aurantia* is easily identifiable as it usually flies close to the ground, and has bright orange fur.

Yellow-bellied Sheath-tailed Bat

Orange Leaf-nosed Bat

Ghost Bat

Rodents

Like the microbats, many of the rodents found in the Top End are difficult to identify unless they are examined closely. They are often seen briefly, scurrying across roads at night or rummaging around in the undergrowth, but except for a few species you will be unlikely to identify them from such a brief glimpse. This is unfortunate, because the group includes some of the Top End's most interesting creatures, with a diverse range of natural histories.

Rodents first evolved in south Asia around 15 million years ago (mya), but do not appear in the Australian fossil record until around 5 mya. They subsequently evolved into a number of species that are endemic to Australia, and are represented in the Top End by species like the **Black-footed Tree Rat** *Mesembriomys gouldii*, **Water Rat** *Hydromys chrysogaster* and **Common Rock Rat** *Zyzomys argurus*; species in this group are called 'old endemics'. Around 1 mya rodents from the worldwide genus *Rattus* first reached Australia and subsequently evolved into several native species, including **Dusky Rat** *Rattus colletti* and **Pale Field Rat** *Rattus tunneyi* which are both found in the Top End; species in this group are called 'new endemics'. The **Black Rat** *Rattus rattus*, common in urban areas that was introduced with the arrival of Europeans, is also from this genus. Through these processes of invasion and evolution, Australia has developed one of the most diverse rodent faunas in the world.

Some Top End rodents have particularly interesting natural histories. The region's most spectacular rodent is the **Black-footed Tree Rat**, one of the largest rats in Australia at around 70 cm long. They are dark above and white below, and have a long dark tail with a white tip. Found in mature savanna woodlands with a dense understorey, they are mostly arboreal (live in trees), sheltering in tree hollows during the day and emerging at night to feed on fruits and seeds. The **Water Rat** is another interesting 'old endemic' which has evolved an aquatic lifestyle. They live in burrows close to creeks or rivers and can swim well, feeding on fish, molluscs, frogs, and even small birds and mammals. Elsewhere in the Top End the rock rats have evolved to live exclusively among the rocky escarpments of Arnhem Land, while the **Kakadu Pebble Mouse** *Pseudomys calabyi*, found only in the Top End, builds small piles of pebbles placed on top of its burrow entrance.

Common Rock Rat

Perhaps the most interesting rodent in the Top End, is one of the 'new endemics', the **Dusky Rat** *Rattus colletti*. Occurring only in the Top End, this species is found on the floodplains of large rivers that are flooded for several months of the year during the wet-season. In response, they migrate to surrounding higher ground, returning to the floodplains as these dry out, and feeding on the vegetation which has sprouted in response to the rains. Incredibly, Water Pythons (*page 243*), one of the rat's major predators, follow them on this migration – one of very few known examples of terrestrial reptile migration.

Black-footed Tree Rat
Water Rat

☐ **Dingo** *Canis dingo* 43–51 in | 1·1–1·3 m

The only wild dog found in Australia, Dingoes are descendants of domesticated Asian dogs. They were first brought to Australia around 3,000–5,000 years ago, probably by Asian fishermen, and since their arrival have spread throughout Australia, becoming an important part of the landscape and Aboriginal culture. Dingoes form loose packs that maintain large territories, and within these territories they may roam as a pack or spend much of their time alone, coming together occasionally to hunt and breed. In the Top End their primary prey is the Agile Wallaby, but they will feed on Magpie Geese, rodents or feral pigs if they are abundant. Dingoes are unique among dog species, only breeding once a year, usually in the mid dry-season. Females maintain a den where they give birth to their pups, keeping them there for about two months before moving them to other dens within their territory. After about four months the pups can survive on their own, and they may remain as part of their natal pack, or disperse to join other packs. One problem facing Dingoes is the degree of hybridization that is occurring with domestic dogs. It is probably true that there are very few 'pure' Dingoes left in the wild. Wild animals are generally quite wary, making them difficult to see, although in some areas they have become accustomed to human activity and will even enter campsites looking for handouts. It is important never to feed Dingoes, as they can become pests.

Where to find Kakadu NP is probably the most reliable place to find Dingoes, but they are just as likely to be seen almost anywhere across the Top End.

☐ **Water Buffalo** *Bubalus bubalis* 98–118 in | 2½–3·0 m

Found across the northern Top End, particularly around floodplains but also in woodland close to permanent water sources, Water Buffalo are an environmental pest responsible for much disturbance in the delicate habitats where they occur. An introduced species, around 1830 about 80 buffalo were brought by Europeans to the first Top End settlements on Melville Island and the Cobourg Peninsula (200 km northeast of Darwin) as a source of meat. Conditions proved too harsh for the settlers, and these colonies were abandoned, but the buffalo remained behind, subsequently colonizing the floodplains and woodlands of the northern Top End. By the mid-1980s the population was nearly 350,000 animals. This large population was responsible for increased soil erosion and destruction of wetland vegetation, having a profound negative effect on native fish, crocodiles, other reptiles and birds. Once the severity of this impact was recognized, a concerted program to eradicate Water Buffalo began in the 1980s. It was particularly effective in Kakadu NP, where numbers were reduced from an estimated 20,000, to fewer than 250 animals in 1996. Populations are now managed carefully, in close consultation with indigenous land managers who have come to rely on the buffalo as a source of meat and income.

Where to find Often seen on the Yellow Water cruise in Kakadu NP.

Reptiles and Amphibians

The warm, arid conditions across much of Australia creates ideal conditions for reptiles. It is therefore no great surprise that the continent supports one of the most diverse reptile faunas in the world. The level of diversity is particularly apparent in the Top End, where reptiles can be found in almost every habitat, from the coast, through the wetlands and floodplains, to the extensive savanna woodlands and rocky escarpments. Indeed, some of Australia's most spectacular reptiles are easily seen in the region, including the world's largest living reptile, the Estuarine Crocodile, and the impressive Frilled Lizard. Although many species are secretive, the intrepid and patient explorer will be rewarded with experiences of some fantastic creatures found nowhere else on earth.

Marbled Velvet Gecko

Green Tree Snake

Estuarine Crocodile

Frilled Lizard

Frogs are the only type of amphibian that occur in Australia, and have one absolutely essential requirement to survive – water. They have permeable skin through which they can lose large amounts of the water from their bodies, and as a consequence must take steps to remain moist and avoid dehydration. In an arid climate like inland Australia, or in places like the Top End where there is no rain for much of the year, frogs have adopted a number of strategies that help them to overcome this problem.

Northern Spadefoot Toad

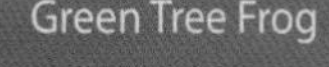

Green Tree Frog

☐ **Estuarine Crocodile** *Crocodylus porosus* 118–275 in | 3·0–7·0 m

Perhaps the Top End's most famous residents, Estuarine or 'Saltwater' Crocodiles are the largest living reptiles and relics of prehistoric times, having changed little since the time of the dinosaurs. Despite their name, 'Salties' are equally at home in freshwater: they are found in coastal areas around the northern Top End but also penetrate far inland on large rivers such as the Alligator River complex, and the Katherine and Victoria Rivers. Extremely efficient aquatic predators, they feed mostly on crabs, fish, reptiles and birds, although large individuals are capable of taking mammals up to the size of buffalo. These large crocodiles are believed to deliver the greatest bite force of any living animal! Hunting by stealth, crocodiles stay submerged and approach their prey underwater before lunging suddenly from the water to seize the victim. Large prey is dragged back into the water and drowned by repeatedly rolling – the famous 'death roll' – and may be stored by wedging it under a submerged log or snag to be eaten later. Male Estuarine Crocodiles are much larger than females, and maintain a particular stretch of river as their territory, aggressively excluding other males. Fights between neighbouring males are serious events, and may lead to injury or even death; it is not uncommon to see large males missing legs or sections of tail as a result of these encounters. The male mates with the females in his section of river, which then build a nest of sand and vegetation on the riverbank that they defend vigorously. Each female lays around 50 eggs which take 2–3 months to hatch. Amazingly, the temperature of the nest determines the sex of the young crocodiles that hatch from the eggs: males being produced in warmer conditions and

females if the nest temperature is cooler. Estuarine Crocodiles have previously been hunted close to extinction, and prior to their protection in 1971 the Top End population was believed to be only about 3,000 animals. It has since recovered, and by the mid-1990s was believed to be around 70,000. Always remember that these fearsome predators may occur in any of the Top End's major waterways, saltwater or freshwater, particularly after the wet-season. NEVER go swimming in a waterhole with signs warning of Estuarine Crocodiles.

Where to find Easily the best place to see Estuarine Crocodiles is on the Yellow Water cruise in Kakadu NP, where you can see them quite safely within metres of your boat.

Where to find Sometimes seen in Katherine Gorge, you can also try searching for this crocodile from the old bridge at Victoria River Crossing. Timber Creek is another good spot, where the owners of the caravan park feed them daily in the small creek behind the park.

☐ Freshwater Crocodile *Crocodylus johnstoni* 79–118 in | 2·0–3·0 m

Smaller than the Estuarine Crocodile (*page 222*), the Freshwater Crocodile has a more slender snout. 'Freshies' are found in most rivers, creeks, lagoons and wetlands across the Top End, and are more likely to occur further inland than Estuarine Crocodiles. Despite the name they are sometimes found in brackish water, although they avoid sections of rivers near the coast. Stealthy predators, they eat insects, fish and small animals, with larger individuals able to take a prey up to the size of a small wallaby. Breeding in the dry-season, females lay their eggs around August. Their clutches are small, with a maximum of 20 eggs being laid in a riverside nest which is not guarded; they hatch after 2–3 months. At this time the female will return to the nest to help the young dig their way out of the nest, and carry them to the water. Unfortunately, the chances of survival for baby crocodiles are slim, with most being eaten by fish, turtles or birds. Freshwater Crocodiles are much more difficult to see than Estuarine Crocodiles as they tend to be much shyer, moving off quietly at the first sign of disturbance. However, with patience and a keen eye, they can often be seen along quieter sections of river.

☐ Northern Long-necked Turtle *Chelodina rugosa* 14½ in | 36 cm

This species has a short, thick neck and large head, with a prominent horizontal yellow stripe through the eye. It is found in the northwestern Top End, where it prefers deep, still water such as permanent billabongs and slow-flowing rivers, and is usually seen floating in the water with just its nose and eyes breaking the surface. They have a varied diet, reportedly eating anything from freshwater mussels to fruit.

Where to find A good place to look is from the small bridge over the lagoon at Howard Springs near Darwin.

☐ Northern Long-necked Turtle *Chelodina rugosa* 14½ in | 36 cm

Found across the Top End, this turtle has a long, thick neck and a large head. It prefers still water such as billabongs and swamps, and also large, slow-flowing rivers. Feeding mostly on small aquatic vertebrates such as tadpoles, frogs and fish, during the wet-season they gorge themselves, forming large fat reserves. During the dry-season, when the wetlands dry out and the turtles are unable to feed, they lie buried in the mud, living on their fat reserves. Once the wet-season arrives and the wetlands fill up, they emerge to begin feeding again.

Where to find Look for this turtle around any wetland, particularly in Kakadu NP.

☐ Pig-nosed Turtle

Carettochelys insculpta 26 in | 65 cm

Where to find Occasionally seen in Kakadu NP.

This very large turtle has flippers rather than 'feet', and a small head with an unusual prominent snout. It is only known from four major drainages in the Top End: the Daly, East Alligator, South Alligator and Victoria Rivers, preferring large, still bodies of water with scattered submerged snags, trees and overhanging banks. It is primarily vegetarian, eating aquatic weeds, fruits and leaves, but will occasionally take insects and molluscs. It has a curious breeding biology; females lay two clutches of eggs in a season, but only breed every second year. The meat and eggs of this distinctive turtle were once an important food source for the indigenous people of the Top End.

☐ Northern Snapping Turtle

Elseya dentata 14 in | 34 cm

Where to find Regularly seen on the Yellow Water cruise in Kakadu NP.

This species has an oval-shaped and quite broad carapace (shell), and the powerful head and neck are dark grey above and pale yellowish below. It is found across the Top End, mostly on larger waterways such as rivers and billabongs. Despite the fearsome name, snapping turtles are mostly herbivorous, eating leaves, fruits and roots, but may take fish and molluscs.

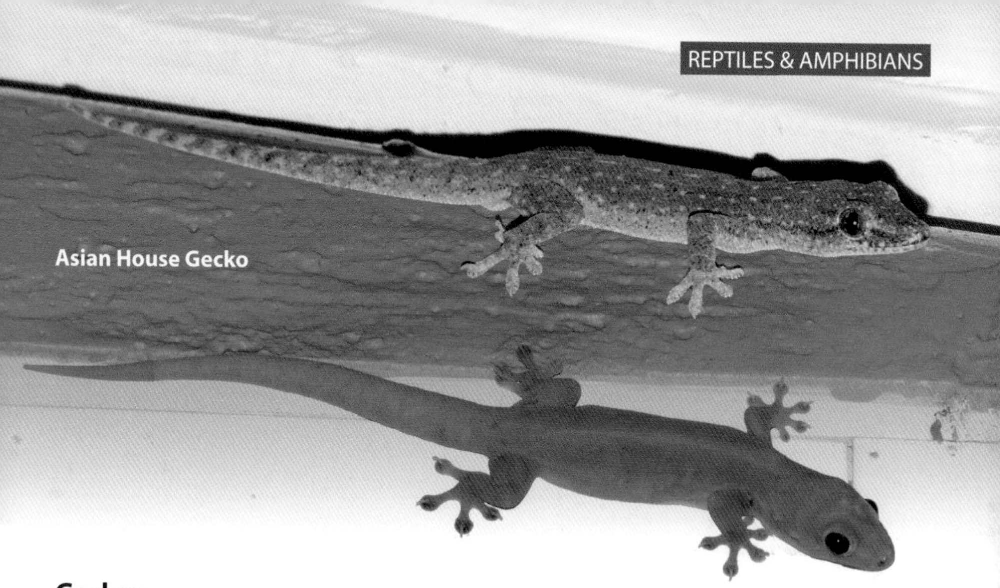

Asian House Gecko

Geckos

Geckos are a widespread group of lizards that are found right across the world, and they are particularly well represented in the Top End. They can be found in most habitats, from woodlands to rocky escarpments and are primarily nocturnal, sheltering under rocks, in caves or in tree hollows by day and emerging at night to feed. Your best chance of seeing most geckos is to explore likely areas at night with a spotlight, as they have a very faint reddish eye-shine which you may be able to see. Once you have spotted one, approach it slowly. A pair of binoculars will help you get a good look before it scurries away. Geckos are commonly thought to be able to cling upside-down to smooth surfaces by using small suction pads on their feet, but in fact their ability to do this is down to static electricity. The underside of each toe is covered with small hair-like projections called setae, each seta branching into many tiny spatulae. Static electricity bonds these spatulae to the surface, and although the force for each one is minuscule, each foot may have millions, providing enough force to bear the gecko's weight. Interestingly, some populations of geckos are parthenogenetic, meaning females are able to produce viable eggs without being fertilized by a male. Geckos born this way are always female, and so populations where this occurs are often all female. Surprisingly, parthenogenesis is not an uncommon phenomenon in the reptile world.

☐ Asian House Gecko

Hemidactylus frenatus 5 in | 12 cm

Where to find Easily found anywhere with buildings.

The first sign that these entertaining little lizards are around is usually their call – a loud "*chk…chk…chk*" often heard around houses and buildings at night and sometimes during the day. They spend the day hiding in dark nooks and crannies, behind paintings, bookshelves and in cupboards, emerging at night to scurry across walls and ceilings hunting small insects. They are slim geckos, with an extremely variable colouration, ranging from almost white to dark grey, often with darker mottling, and can change colour depending on the background.

☐ Bynoe's Gecko *Heteronotia binoei* 4½ in | 11 cm

One of Australia's most widespread reptiles, these little geckos are also one of the most common. They are terrestrial and spend the day hiding beneath leaf-litter, logs, rocks or other cover, emerging at night to forage for small insects and other invertebrates. They can be confusing to identify; like many reptiles they are extremely variable in pattern and colouration, but are often dark brown with light bands. Also keep an eye out for the similar **Banded Prickly Gecko** *Heteronotia planiceps*, which is found on rocky escarpments in Kakadu NP and also in the Victoria River Region; it is a little more slender and has more sharply defined bands.

Where to find Can be found anywhere, particularly on warm nights, but usually in rocky areas with plenty of leaf-litter.

Bynoe's Gecko

Banded Prickly Gecko

Where to find Widespread, but the Escarpment Walk at Victoria River Crossing is a good place to look.

☐ Northern Spotted Dtella

Gehyra nana 4½ in | 11 cm

This slender greyish or grey-brown gecko is usually covered in small black and white spots. It is fairly common on rocky outcrops or escarpments but can be difficult to see well as it scurries quickly out of sight when disturbed, often squashing itself into seemingly impossible cracks or holes in the rock as it escapes.

☐ Northern Dtella

Gehyra australis 6½ in | 16 cm

Quite common in the Top End, this chunky gecko is variably mottled greyish or grey-brown above, without any clear pattern, and is pale below. Usually found scurrying up and down the trunks of trees, and sometimes under eaves or along the walls of houses, it looks quite 'muscular'. Like all geckos and many skinks, when attacked it is able to shed its tail, which then writhes around on the ground, distracting predators and allowing the now tail-less gecko to escape. This is a last resort though, as the tail often stores much of the animal's fat reserves.

Where to find Wherever there are trees, and often around buildings.

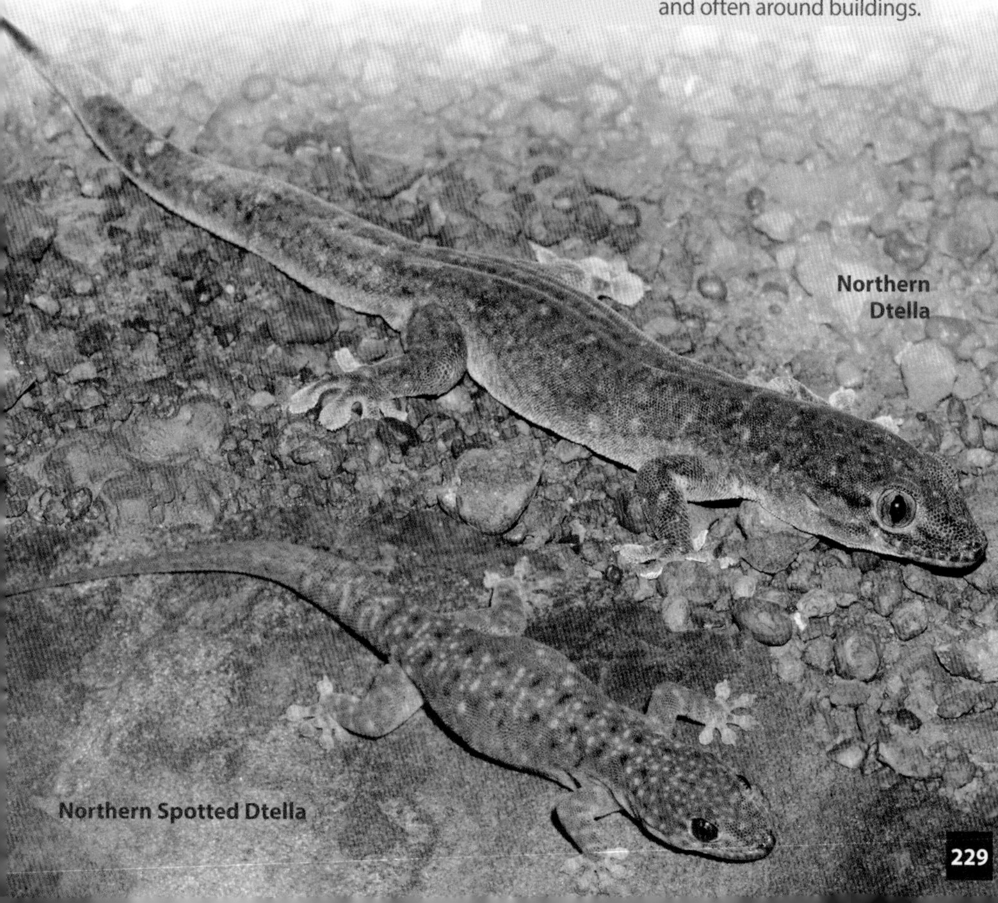

Northern Dtella

Northern Spotted Dtella

■ Northern Spiny-tailed Gecko

Strophurus ciliaris 6½in | 16cm

Named after the rows of tiny spines on their tail, these geckos are common in drier habitats right across the Top End. They are mainly arboreal, spending most of their time in the canopy of trees or shrubs but sheltering under bark or in tree hollows during the day. If threatened, they have the curious ability to exude a sticky brown substance from their tails. These geckos are extremely variable in colour and pattern, with some individuals being plain greyish or brownish and others having white, orange or grey blotches or spots on their body. As well as the spiny tail, they also usually have a couple of small spines just above the eye. The eye itself is incredibly patterned; if you are lucky enough to get a close view of one, look closely at the maze of orange, white and black in the iris – one of nature's most spectacular creations.

Where to find One of the most widespread geckos, which is active almost anywhere at night.

Marbled Velvet Gecko *Oedura marmorata* 7½ in | 19 cm

A fairly large gecko, adults are shades of purplish-brown, variably banded and spotted with yellow; the young are black with bright yellow bands. Widespread across Australia, in the Top End this beautiful gecko is found mostly around rocky outcrops or escarpments. It spends the day sheltering in crevices or beneath slabs of rock, emerging at night to forage for insects and spiders on exposed rock faces.

Where to find Rocky habitats anywhere; though a good place to look is the top of the escarpment walk near Victoria River Roadhouse.

Northern Knob-tailed Gecko
Nephrurus sheai 5 in | 13 cm

Where to find In woodlands associated with rocky outcrops and escarpments.

These geckos have a very large, triangular head, huge eyes, and a tiny tail with a small knob at the end. It is not known why the tail is shrivelled-up, or the purpose of the tiny knob at the end, although it is believed to be quite sensitive to touch. Knob-tailed Geckos also have an unusual display when threatened: they raise themselves high on their scrawny legs, like a miniature reptilian horse, and sway gently, lunging at the threat and sometimes making a small barking sound!

■ Giant Cave Gecko *Pseudothecadactylus lindneri* 7 in | 18 cm)

This very large gecko has a large, triangular head and a purplish-brown body with a number of patchy yellowish bands. The tail is more sharply banded yellow, although regenerated tails are plain. It is restricted to the sandstone escarpments of western Arnhem Land and spends most of the day sheltering deep in caves and crevices, emerging at night to hunt insects, spiders and other invertebrates, and sometimes even other geckos. This gecko is representative of one of the most ancient vertebrate lineages in Australia, evolving from an ancestor that diverged from its nearest living relatives more than 40 million years ago. This makes it more ancient than many other typically 'Australian' vertebrates such as kangaroos. Once widespread, as the climate across northern Australia became drier, it survived in only a few small pockets such as the Arnhem Land escarpments. Interestingly, there are many differences between populations of this gecko even within its small range, differences that are not apparent to us but can be recognized through laboratory techniques such as genetic analysis. These differences indicate that even populations which might be relatively close to each other as the crow flies, are actually genetically isolated.

Where to find Around some of the sandstone escarpments in Kakadu NP, including Nourlangie and Gunlom (but be aware that access to some of these areas is not permitted at night). The best time to look is on warm evenings during the wet-season.

☐ **Burton's Snake-lizard** *Lialis burtonis* 10 in | 25 cm

Where to find Possible in most habitats across the region.

Australia's most widespread reptile, this is the most commonly encountered member of a fascinating group – the flap-footed, or legless lizards. They are often confused with snakes, but are actually lizards that over time have lost their legs. Some species still have residual flaps of skin on the sides of their bodies, the remnants of legs, and many still have tiny bones inside their bodies where their legs used to be, but none of them have useable legs. This species is distinctive because of its sharp, pointed, triangular head. It is quite variable in colouration, but most are varying shades of brown or grey, often with a white stripe through the mouth. Most snake-lizards feed on insects and spiders, but Burton's Snake-lizard is unique because it feeds almost entirely on other lizards, usually small skinks (see *page 234*). It has a curious foraging behaviour, being a 'sit-and-wait' predator that spends most of its time buried in leaf-litter waiting for prey to come close, sometimes waving its tail, probably to lure or distract the prey, before pouncing. It has an unusual mouth with a hinge across the upper jaw, like a pair of multi-grips, which allows it to grasp large prey.

Skinks

If you have ever been walking along a trail or in the garden, and wondered what the small lizards are you see scurrying off the path in front of you, the chances are they are a species of skink. Skinks are one of the most varied groups of vertebrates, with nearly 450 species recognized in Australia. They have evolved to exploit almost every habitat in the country, from moist rainforests to hot, dry deserts, and from the tropical north of the Northern Territory, to the freezing southern tip of Tasmania. They have also evolved an incredibly wide range of forms, from animals which are practically legless and spend their lives underground, to large species like the blue-tongues.

Skinks are very well represented in the Top End and are found in most habitats, from the gardens of suburban Darwin,

Eastern Striped Skink
Ctenotus robustus

Top End (or Northern) Fire-tailed Skink
Morethia storri

to the rocky escarpments of Arnhem Land. Unfortunately, the wide variety of forms and the difficulty of getting close enough to see them, makes many of them very difficult to identify. A good technique if you are able, is to take a digital photo at full zoom, then use your digital screen or computer to zoom right in and get a better view. However, in some cases identification is not possible without being able to hold the animal and examine it with a magnifying glass! Still, there are a handful of species that are more common than others, and some pictures of these are included here. Ultimately though, you will probably have to satisfy yourself that what you have just seen is a skink of some sort, and that you will need a better look next time!

Slender Rainbow-skink
Carlia gracilis

Red-sided Rainbow-skink
Carlia rufilatus

☐ Common Blue-tongue *Tiliqua scinciodes* 14 in | 35 cm

Widespread across Australia, these well-known lizards are frequently found in gardens, often taking up residence under houses, hot water systems or wood piles, and emerging to bask in the sun on open areas such as paths. They are diurnal, meaning they are most active during the day, and shelter in hollow logs or under other cover at night. Large, stout-bodied lizards with short legs and a large head, they are generally quite slow-moving, although can put on a turn of speed for short distances if required. In the Top End they are usually rich brown above with irregular bands of black and pale brown, and are best known for their threat display, puffing up their body and gaping, showing off their bright blue tongue. They feed on a variety of vegetation, fruit, snails, insects and small vertebrates, and have even been found eating dog food! Blue-tongues are viviparous, meaning females give birth to live young, sometimes as many as 25 in one go!

Where to find Possible in most habitats, including suburban areas, and often seen slowly crawling across the road.

Gilbert's Dragon

Lophognathus gilberti 16 in | 40 cm

This slender lizard has a very long tail, up to three times the length of its body. Males are dark, almost black, below and have a prominent white stripe running from the mouth back along the side of the body, and often have a rusty coloured tail. Females are more slender and browner, with a pale throat and two parallel pale lines down the back. Confusingly, another species, **Swamplands Lashtail** *Gowidon temporalis*, also occurs in the Top End and is so similar to Gilbert's Dragon that only experienced reptile experts can tell the two species apart. They do, however, have different habitat preferences; Gilbert's Dragon is found in drier woodlands, particularly farther south in the region, whereas Swamplands Lashtail favours wetter habitats, such as the margins of creeks and rivers. The lashtail is quite common in suburban Darwin, where it is often seen sunning itself on trees and fence posts in parks and gardens.

Gilbert's Dragon
Female

Swamplands Lashtail
Male

Where to find **Gilbert's Dragon** can be seen almost anywhere in the Top End. Good places to see **Swamplands Lashtail** include East Point Reserve in Darwin.

☐ Frilled Lizard

Chlamydosaurus kingii

35½in | 90cm

An iconic species, the 'frill-neck' certainly has an eye-catching threat display. When startled, the large cape of loose skin around the neck is flared out while the animal gapes, showing the bright yellow inside to its mouth. Do not expect to see this display, though, as when threatened they are more likely to sprint away on their hind legs and bolt up a tree. They feed mainly on termites, and this diet affects their behaviour throughout the year. In the dry-season when food is scarce they spend their time high in trees, moving very little. However, as the wet-season arrives they become more active, spending much of their time clinging to trunks searching for food on the ground which they will run down to capture before climbing back to their vantage point. While clinging to trunks they are difficult to see, shuffling around to keep away from any disturbance. Your best chance of seeing one is when they are crossing the road. If you stay in the car and approach slowly they will often sit there watching you, but open the door and off they shoot!

Where to find Open woodlands, parks and gardens anywhere in the region.

☐ Sand Goanna *Varanus gouldii* 63 in | 1·6 m

These very large, slender lizards have a long, pointed snout, powerful legs, long claws and a very long tail. They are quite variable in colour but are mainly yellowish with dark speckling, mottling and spotting. Known as 'monitors' elsewhere in the world, 'goanna' is a peculiarly Australian term given to these large lizards. This name is not of Aboriginal origin as often believed, but is apparently a corruption of 'iguana', the name given to several large lizards found in Central and South America. Sand Goannas are widespread and quite common across the Top End and indeed right across Australia. They shelter in burrows or hollow logs, and are active during the day, fossicking on the ground and flicking their long forked tongue, searching for just about anything they can eat, from insects and small mammals, to birds' eggs and sometimes even carrion. The very similar **Yellow-spotted Monitor** *Varanus panoptes* (not illustrated) is also found in the Top End, often around water, but only experienced herpetologists can tell these species apart in the field. Like many large lizards they are most active in the wet-season when prey is abundant, spending most of the dry-season sheltering and conserving energy while food is scarce.

Where to find Often seen crossing roads, mostly in drier, more open habitats.

☐ Merten's Water Monitor *Varanus mertensi* 43½ in | 1·1 m

These semi-aquatic lizards have a long pointed snout, yellowish throat and belly, and a dark grey body with scattered yellow spots, each surrounded by a dark ring. They are common residents around water across the Top End, although the arrival of Cane Toads (*page 263*) is thought to have resulted in population declines in some areas. Well adapted to their aquatic lifestyle, they have a strongly compressed tail, like a crocodile's, that allows them to swim very well. They also have nostrils on top of their snout to help them breath while staying mostly submerged. Often seen basking on rocks near waterholes or on the edges of wetlands, they quickly launch themselves into the water if they are disturbed. They are quite shy and can be difficult to see, but around some busy waterholes may become used to people and allow much closer approach. In Kakadu NP they feed mostly on freshwater crabs, but will eat almost anything they can find, including turtles' eggs, frogs and other small vertebrates.

Where to find Possible anywhere across the region, but always around permanent waterholes, wetlands and sometimes rivers and creeks.

☐ Children's Python *Antaresia childreni* 40 in | 1·0 m

Named after the British zoologist John George Children, these snakes are one of the world's smallest pythons, and although they may grow to one metre in length, are usually much smaller. Pale to dark brown and covered in darker brown blotches, they are extremely variable, with some animals strongly patterened and others appearing almost plain. Because of their small size and docile nature they are popular pets, easy to keep in captivity and becoming quite used to being handled. They spend the day sheltering in caves, hollows logs or disused burrows, emerging at night to forage for small mammals, birds and reptiles. They spend most of their time on the ground, but are agile climbers, and will ascend trees in search of food. In some areas they are known to take up positions on rocky ledges at the entrances to caves where they try to catch bats emerging from their roost at dusk. These snakes are quite common across the Top End, and are found in most dry habitats, although they show a definite preference for rocky areas. Like all pythons, Children's Pythons are oviparous, meaning they lay eggs. However, unlike most snakes that reproduce this way, female pythons care for their clutch, incubating the eggs until they hatch and the hatchlings disperse.

Where to find Drier habitats such as savannas, particularly around rocky outcrops or escarpments, but most often encountered crossing roads at night.

☐ Olive Python *Liasis olivaceus* 158 in | 4·0 m

The largest snake in the Top End and second largest in Australia (the largest is the **Amethystine Python** *Morelia amethistina* of northeastern Queensland), there are reports of these snakes reaching more than six metres in length! They are variable in colour, but usually pale brown above with a faint sheen, and pale cream below. Preying mostly on mammals, large individuals can eat animals up to the size of a wallaby. Like most pythons they kill their prey by constriction, capturing the animal and coiling their body around it, slowly squeezing tighter and tighter until the victim has suffocated. They then eat their prey whole, starting with the head. Python skulls are specially adapted to allow them to eat such large prey, with the jaws held together with flexible ligaments, and necks that can stretch enormously to accommodate prey many times larger than the snake might otherwise be able to swallow. After having eaten such a large meal, it is important for the snake to maintain a high body temperature to aid digestion, and they usually achieve this by basking in the sun. If it is too cold the food will not digest and may rot inside the snake, poisoning it. These snakes can be found anywhere across the Top End, and are most often seen crossing roads at night.

Where to find Search in drier areas, particularly around rocky outcrops or escarpments.

☐ Water Python *Liasis mackloti* 100 in | 2·3 m

Compared to the similar Olive Python, Water Pythons have a much narrower head and a bright yellow belly. The dark, glossy brown body also has an iridescent sheen. Although most common close to water, Water Pythons can wander far, particularly in the wetter parts of the northern Top End. However, farther south, such as in the drier areas around Katherine, they are only found around wetlands, waterholes and rivers. In some areas of the Top End, such as the wide Alligator River floodplains of northern Kakadu NP, they are one of the major predators. The populations here are well-studied and exhibit one of the few known examples of reptile migration. Their main prey is the Dusky Rat (see *page 216*), which moves to higher ground during the wet-season, but returns to the floodplains during the dry-season. The pythons follow the rats, annually moving distances of up to 10 km to follow their prey. In some years, a combination of extreme wet-seasons and an extended dry-season can cause the rat population to collapse, and without a reliable source of prey, many pythons die of starvation. Elsewhere, where the Dusky Rat is not the main food source, they will eat other mammals, birds and even small crocodiles. Water Pythons are most often encountered crossing roads at night.

Where to find Fogg Dam near Darwin is a good place to look. It is also often seen along the Arnhem Highway west of Jabiru where the highway traverses the floodplains of the Alligator Rivers.

☐ Black-headed Python *Aspidites melanocephalus* 103 in | 2·6 m

These distinctive large pythons have a yellowish to light brown body with numerous darker brown bands, and a jet black head and neck. Although they look menacing, like all pythons they are not venomous. They can be a bit grouchy, and while their first reaction to a threat is to move off, if harassed they may hiss and lunge, but rarely bite. Although occasionally feeding on mammals and birds, they are unusual among the pythons in that they feed almost exclusively on other reptiles, mostly larger lizards such as dragons. They sometimes eat other snakes, including venomous species, and are apparently immune to their venom. Many pythons have special heat-sensitive pits on the sides of their mouth which they use to detect warm-blooded prey, but these are lacking in Black-headed Pythons, probably because their reptile prey is mostly cold-blooded. They spend much of their time sheltering in burrows, hollow logs and termite mounds, where they partially bury themselves in loose soil or leaf-litter, often with the head sticking out. It is thought that the black head is an adaptation that enables them to warm up; black objects absorb heat best and by holding just their head in the open while their body remains buried, they can still benefit from the heat of the sun.

Where to find Most common in drier woodlands and often seen crossing roads at night.

☐ Oenpelli Python *Morelia oenpelliensis* 158 in | 4·0 m

These very large pythons have a big head and a pale brown body with rows of prominent brown blotches. They are found only in the rocky escarpments of western Arnhem Land. Although the local aboriginal people probably knew of the existence of these snakes for thousands of years, they remained unknown to 'science' until the late 1970s due to their very restricted range and specific habitat preferences. They are believed to feed mainly on mammals, and being a large snake can take prey up to the size of a rock-wallaby; however, possums, flying-foxes and birds are more commonly taken. Unfortunately their rarity makes them a target for illegal collection, although the Northern Territory government is taking steps to prevent this by trying to establish a captive-breeding population. It is very unlikely that you will see one of these snakes as they spend the day sheltering in crevices, caves and tree hollows, emerging at night to hunt among the rocky escarpments and nearby rocky outliers. Also, the areas where they are found are mostly inaccessible without special permits – but there is always a chance!

Where to find Sometimes seen around Nourlangie Rock in Kakadu NP, but remember that access to some of the places where this snake is found is not permitted at night.

☐ Carpet Python *Morelia spilota* 118 in | 3·0 m

This large, beautifully patterned python is one of Australia's best-known snakes and found in most habitats across the Top End. It is commonly seen in suburban areas where the dark lofts of houses provide perfect shelter, and where the numerous possums, rats and birds attracted to the fruit trees in gardens provide plenty of food. Many a Darwin resident has been startled to find that they have unwittingly played host to a Carpet Python when the electrician climbs into the roof to install a new light, years since their last visit, and emerges with a handful of snake skins each several feet long! This species is widespread throughout Australia, with populations from different areas each having a distinctive pattern. The animals from the Top End are particularly sought after by reptile enthusiasts and are well established in captivity – breeders often calling them 'Darwin' Carpet Pythons. Like most pythons it is nocturnal, and this species is also mostly arboreal, feeding in trees primarily on mammals and birds, with large pythons able to take prey up to the size of a possum.

Where to find Just about anywhere, and common in suburban Darwin.

Brown Tree Snake *Boiga irregularis* 79 in | 2·0 m

WEAKLY VENOMOUS

This distinctive, slender snake has a disproportionately large head in relation to the width of the body, and very large eyes. Animals from the Top End are much more boldy patterned than their counterparts from eastern Australia, having a pale yellow body with numerous dark brown bands – earning it the name 'Night Tiger'. Found across northern and eastern Australia, this species is weakly venomous but not considered dangerous, with only small fangs at the rear of the mouth. It is nocturnal, sheltering by day in tree hollows, caves or the ceilings of buildings, and emerging at night to hunt birds and small mammals. As the name suggests, it is mostly arboreal (found in trees), although it is often found moving about on the ground. Like Children's Python (*page 241*), it sometimes positions itself at the entrances to caves at dusk, striking swiftly at bats as they emerge from their roosts. The slender body allows it to squeeze through the smallest of gaps, including between the bars of a bird cage, and many a poor pet owner has awoken in the morning to find their pet Budgerigar replaced by a well-fed Brown Tree Snake!

Where to find Almost anywhere across the region, including suburban gardens, and often seen crossing roads at night.

☐ **Green Tree Snake** *Dendrelaphis punctulata* 51 in | 1·3 m

(Golden Tree Snake)

Found across northern and eastern Australia, these snakes vary greatly in colour across their range, and it is the animals from southern Queensland that give this species the name 'Green' Tree Snake. In the Top End most animals are a beautiful golden colour, earning them the common name 'Golden' Tree Snake. They also have a bright yellow belly, and usually a greyish head. When threatened, they flatten their neck, and if you are close enough you may see the beautiful pale blue skin between the scales. But, be careful – although considered harmless to humans due to the small fangs at the rear of their mouth and weak venom, their bite sometimes causes a mild local reaction. These snakes appear to do particularly well in suburban areas, often being seen in gardens, or entering homes. They are mostly diurnal (active during the day), and feed mainly on frogs, lizards and occasionally birds. Spending most of their time in trees, they are very agile, moving quickly and effortlessly through the outer branches. They are occasionally seen on the ground, often crossing roads, and if startled can slither away very swiftly. When not active, they shelter in dark spaces, including tree hollows or caves, and sometimes the roof spaces of houses.

Where to find Commonly seen crossing roads, and regularly encountered on the Yellow Water cruise in Kakadu NP, usually basking in a tree over the water.

Although methods of measuring how venomous a snake is vary, the Taipan is generally recognized as being the third or fourth most venomous snake of all, its poison being about 40 times more potent than that of the Eastern Diamondback Rattlesnake *Crotalus adamanteus* of North America. Its close cousin the Inland Taipan *Oxyuranus microlepidotus*, found in the deserts of southwestern Queensland is recognized as the most venomous snake in the world, its venom being twice as potent again!

☐ **Taipan** *Oxyuranus scutellatus* 79 in | 2·0 m **EXTREMELY VENOMOUS**

This long, slender species, can vary in colouration from pale to quite dark brown, although the small head is usually paler than the body. It is found in wetter areas across the northern Top End, mostly in grassy woodlands. Australia has a reputation as being home to some of the world's most venomous snakes, and this species sits firmly in the top five. Like most snakes, its first reaction is to flee, and this very fast snake can get away quickly. However, if it is cornered it will not hesitate to bite. Needless to say, it is EXTREMELY DANGEROUS, and should be left alone. Mostly active during the day, in the Top End the warm climate means it is also often active at night. It is unusual among Australia's venomous snakes in eating only mammals, with rats the most common prey. A rat can easily kill a snake with its powerful bite, so this snake has evolved a unique hunting strategy. It bites its prey, injecting it with a large amount of venom, but releases it immediately to avoid retaliation. So much of the powerful venom is injected, that the victim dies very quickly, allowing the snake to retrieve and swallow it safely.

Where to find Almost anywhere across the region, particularly around grassy areas at night.

ALL THESE SNAKES ARE EXTREMELY VENOMOUS

☐ Northern Brown Snake

Pseudonaja nuchalis 51 in | 1·3 m

This highly variable species may be almost any shade of brown, with speckles, bands or blotches. Occurring across the Top End, this species and its close relatives are often ranked as Australia's second most venomous snakes. Their variability makes them difficult for the inexperienced observer to identify, and you need close views to do so – which is not recommended as they are EXTREMELY DANGEROUS! They are nervous and will strike if cornered, first rearing the front of their body in an elevated 'S'-shape. Given the chance to escape, they will – and this is always the safest option if you come across one.

Where to find Mostly in drier habitats, such as savannas, but also does well in suburban areas.

☐ Northern Death Adder

Acanthopis praelongus 20 in | 50 cm

This small, 'fat' snake ranges in colour from reddish brown to grey, with paler bands. Another EXTREMELY DANGEROUS species, this species is an ambush predator and a fine example of convergent evolution. It shares many similar characteristics with the vipers of Asia and Africa, including a short, squat body and its hunting technique, but is not closely related. Instead, it has evolved many of the same characteristics due to its similar ecology. It spends most of its time partly buried in the leaf-litter curled in a 'C'-shape, with its head close to its tail. After spotting its prey, usually a small mammal or reptile, it begins vigorously wiggling its tail, using it as a lure before striking, with the powerfully neurotoxic venom soon overpowering the victim. Unable to flee quickly, when threatened they sit tight, and if provoked will flatten their body and change colour, the pale bands becoming much more prominent, before finally striking. Previously common on the floodplains of Kakadu NP, these populations have apparently declined since the arrival of Cane Toads (*page 263*).

Where to find Usually seen crossing roads at night, and can be found anywhere; Fogg Dam is a good place to search for them.

☐ Mulga Snake

Pseudechis australis 99 in | 2·5 m

This is one of the largest venomous snakes in the world, and although still EXTREMELY DANGEROUS ranks a little way down the list, at around fifteenth. It is known to inject more venom in a single bite than any other snake, and is also the heaviest of Australia's venomous snakes, with large individuals weighing up to 6 kg. This species is quite common in the Top End, but anecdotal reports suggest it has suffered a decline since the arrival of Cane Toads (*page 263*).

Where to find Regularly seen around floodplains, you may encounter this species crossing the Arnhem Highway at night, where it traverses the floodplains of the Alligator Rivers.

Northern Brown Snake

Northern Death Adder

Mulga Snake

☐ **Keelback** *Tropidonophis mairii* 39 in | 1·0 m

This fairly small snake occasionally grows to one metre in length, but is usually much smaller. It is quite variable in colour, but is usually pale cream below and brownish above, with patterns of darker spots, flecks and bands. Found across the Top End, it always occurs close to water, along creeks or rivers, on floodplains, and around the edges of wetlands. It does well in suburban areas, and is often found in moist gardens. The Keelback is perhaps best known for its resistance to the poison of the Cane Toad (*page 263*), being able to eat them without any ill-effects. Why this species is resistant to the toad's toxins while other Australian snakes are not is a mystery, but unlike many other frog-eating snakes, its populations have not been adversely affected by the arrival of the toads. However, although they are resistant to the toad's toxins, they still avoid eating them and prefer to prey on frogs and occasionally lizards. The Keelback gets its name from the ridge, or keel, on each scale, but this can only be seen if you are able to get very close. These keels often give the appearance of a number of thin parallel lines running the length of the snake. It is not venomous, but as with all snakes it is best not to get too close; snakes can be difficult to identify for the inexperienced observer, and you would not want to make a mistake!

Where to find Quite common on the floodplains of Kakadu NP, and also easy to find at Fogg Dam.

☐ Arafura File Snake *Acrochordus arafurae* 79 in | 2·0 m

These unusual aquatic snakes are non-venomous and found across the northern Top End, always associated with wetlands, waterholes and rivers. Their loose, baggy skin shrouds a muscular body, and they are very capable swimmers; they rarely emerge onto land, where they are slow-moving and practically helpless. They have several other adaptations to their almost exclusively aquatic lifestyle including a low metabolic rate, allowing them to remain submerged for extended periods and eat very little. They spend most of their time buried in mud among snags, tree roots or on riverbanks, and mostly eat fish which they find by searching actively underwater or ambush. In the wet-season they disperse with the floodwaters, but as the dry-season progresses and waterholes dry up, they may congregate in large numbers where water remains. At this time they are an important food source for Aborigines, who search for them in the mud and throw them onto the bank where they are unable to crawl away; they are then collected once the hunt is complete. However, the presence of Estuarine Crocodiles (*page 222*) in many of the same places where file snakes are found makes searching for them a hazardous process!

Where to find This species is common in Kakadu NP, and may be found late in the dry-season when it occurs in small, isolated waterholes.

☐ **Red Tree Frog** *Litoria rubella*
1½ in | 4 cm

One of Australia's most common and widespread frogs, this is another species that does well in and around houses and outbuildings. It seems better able to cope with living in drier areas than many 'tree' frogs, and can be found almost anywhere there is water. Its call is a high-pitched, drawn-out "*reeeep-reeeep-reeeep...*" with an upward inflection at the end of each note. It is a small frog and quite variable in colour, but is usually a reddish shade of brown with a darker stripe through the eye.

Where to find Easily found around houses or gardens after rain.

☐ **Roth's Tree Frog** *Litoria rothi* 2½ in | 5½ cm
('Laughing' Tree Frog)

This medium-sized, long-legged frog is very variable in colour but is usually brownish or greyish and always has black-and-yellow patterning at the back of the thighs. It is found right across the Top End in most habitats, usually close to water. The call is commonly heard around waterholes, a rapid 5–8-note descending "*uh-uh-uh-uh-uh*" which also gives this species its alternative English name.

Where to find Around the edges of waterholes, where they are often perched on a branch close to or over the water.

☐ Green Tree Frog *Litoria caerulea* 4 in | 10 cm

Probably Australia's best-known frog, this species will take up residence almost anywhere there is persistent moisture, and most campground shower blocks have a few in residence. Many visitors to the Top End have flushed the toilet, only to be startled by one of these frogs shooting out from its moist hidey-hole under the lip of the bowl! They are big frogs, with females being much larger than males. They are bright green above and white below, and have a large pad at the end of each of their fingers and toes that helps them cling to smooth surfaces. These frogs do very well around houses and buildings where they shelter in toilet cisterns, behind paintings and bookshelves, and in downpipes. They find plenty of food around the house too, with some animals learning to sit close to outdoor lights at night, feeding on the insects they attract. Their call is well known, and is often given at the onset of rain by males in an effort to attract a mate. It is a deep "*rork-rork-rork...*" that reverberates from inside the drainpipe or tree hollow where the frog is sheltering. Away from buildings they are usually found in wetter habitats or along watercourses, where they shelter by day in tree hollows, emerging at night to feed on insects and other invertebrates. They also occasionally eat small animals, including lizards and sometimes other frogs.

Where to find Common across the region and likely to be found during any short walk outside after dark with a torch.

☐ Northern Dwarf Tree Frog

Litoria bicolor 1 in | 2½ cm

These tiny frogs are usually green, although sometimes brown, with a paler belly; a white stripe runs from below each eye and along the flanks. They are found across the far northern Top End, and spend much of their time in trees or on sedges close to the water. Except for their call, a distinctive high-pitched, staccato and upwardly inflected "*riiii-ik ik-ik*", you would hardly know they were around. Even so, trying to find one, despite being able to hear it calling nearby, is a frustrating experience!

Where to find Search for this frog when they start calling after rain.

☐ Rocket Frog *Litoria nasuta* 2 in | 5 cm

This slender frog has a pointed nose and very long, spindly legs. It is variable in colour, but usually brown or grey with darker blotches and mottling. Aptly named, it is not uncommon having spotted this species for it suddenly to disappear with an enormous leap of several metres. It is found right across the Top End and is quite common in grassy areas close to water. The call is a high-pitched and quickly repeated "*rit-rit-rit…*", each note inflected upwards.

Where to find Search at night, in grass close to the edge of waterholes or wetlands.

☐ Giant Frog *Litoria australis* 4 in | 10 cm

A large, stocky frog with a triangular snout and huge mouth, this species is extremely variable in colour, and can be brown, green or even yellowish, usually with a dark line through each eye. It is a voracious predator that feeds mostly on other frogs and is able to eat prey nearly as big as itself. To overcome the problem of desiccation (dehydration), this frog buries itself for the duration of the dry-season. Around May, as the dry-season sets in, it burrows into the soil and makes a small chamber around 5–10 cm below the surface. As the dry-season continues and the soil continues to dry, it sheds layers of skin, creating a waterproof cocoon that protects it from further water loss. When the rains arrive around October it sheds this cocoon and returns to the surface to feed and breed. This behaviour is not uncommon in Australian frogs that have evolved to survive in a dry climate, and there are some species found in the deserts of central Australia that spend years underground waiting for rain to arrive. The Giant Frog is quite common across the Top End, but is most often seen during the wet-season.

Where to find Open habitats such as paddocks, grassy woodland and wetlands; often found sitting on roads after periods of heavy rain.

Dahl's Aquatic Frog

☐ Dahl's Aquatic Frog *Litoria dahlii* 3 in | 7 cm ▲

This large frog is extremely variable in colouration, and although usually grass-green above, it may be any shade from bright green to brown. The patterning on the back is quite variable too, and may be almost plain or covered with faint brown blotches. As the name suggests, this species is always found near water and can be very common around some of the permanent wetlands across the Top End. Mostly nocturnal, it is usually found sitting among vegetation at the water's edge, but is sometimes seen basking during the day. It feeds on other frogs and occasionally tadpoles and small fish.

Where to find Fogg Dam near Darwin is a good place to look.

☐ Copland's Rock Frog *Litoria coplandi* 1½ in | 3½ cm ▶

This small frog is pale brown above, mottled with darker brown spots and blotches, and very pale brown or white below. It spends the day sheltering in crevices or under rocks, emerging at night to feed around rockpools and waterholes. This species is not as reliant on permanent waterholes as Masked or Rockhole Frogs, and is able to survive around small seepages or drips in the rock face.

Where to find Around rocky waterholes, creeks and escarpments anywhere in the region.

Where to find Look for this species in the waterholes at the top of the Gunlom waterfall in Kakadu NP.

☐ Rockhole Frog *Litoria meiriana*

1 in | 2 cm

These tiny frogs are usually mottled brown and pale brown, sometimes appearing banded on the hind legs, and have small, orange eyes. They are found around rockholes and rocky creeks from Kakadu NP to the southwestern Top End where, in the late dry-season, they can often be seen crowded around the edges of the few remaining waterholes. These frogs are so small that they are able to hop across the water without breaking the surface!

Where to find Waterholes at the top of the Gunlom waterfall in Kakadu NP are a good place to look.

☐ Masked Frog *Litoria personata*

1½ in | 3½ cm

Small and long-legged with a triangular snout, these frogs are dark above and very pale brown or white below, with a dark stripe through the eye. One of only a few species of frog that are restricted to the Top End, they are found only in rocky creeks and waterholes on the escarpments of western Arnhem Land.

Masked Frog

Rockhole Frog

Copland's Rock Frog

Marbled Frog *Limnodynastes convexiusculus* 2½ in | 5½ cm

A small frog with a longish, rounded snout, this species is usually pale brown or grey, and covered in large, green blotches above that often form a symmetrical pattern. It is quite a common frog, generally found in thick grassy areas, including woodland clearings, and often gathering around swamps or any wet, grassy area. It is best detected by its distinctive rapid, high-pitched "*bonk-bonk-bonk…*" call, given at a rate of about two "*bonk*"s per second. If you hear this frog calling, approach slowly – and, if it stops, stand still until it starts again before moving closer.

Where to find Search for this species in grassy woodlands, particularly in the south and southwest of the region.

Where to find Search for this frog in paddocks and grasslands at night, immediately after rain.

Ornate Burrowing Frog
Platyplectrum ornatum 1½ in | 4 cm

This small but robust frog has a prominent blunt snout and relatively large eyes. It is extremely variable in colour and pattern, but often has a line running down the middle of the back. Found across the Top End, usually in open habitats like savannas and grasslands that are seasonally flooded, this is one of the burrowing species that spends the dry-season buried deep underground, emerging during the wet-season to breed and feed.

Where to find You are only likely to come across this species at night during the wet-season, and immediately after rain, when they are often seen crossing the road.

■ Northern Spadefoot Toad *Notaden melanoscaphus* 2 in | 5 cm
('Golfball' Frog)

These small, strange-looking frogs have a large, bulbous body and small, blunt head. They are greyish, greenish or brownish above, usually with large, darker blotches. Found in poorly drained grassy areas, such as open savannas and plains, they are a burrowing species, spending the dry-season deep underground. Unlike other burrowing frogs which remain above ground during the wet-season, Northern Spadefoot Toads bury themselves each day, only emerging at night immediately after rain. When they do come to the surface they congregate around swamps, with males sitting spread-eagled in shallow water and giving their unusual call, a repeated "*woop-woop-woop...*" at a rate of about one note per second. The males all call together in synchrony, and the sound of this chorus can travel some distance. Since they are unable to move quickly on land, when confronted by a predator they puff up their bodies until they look like a small sphere, hence their alternative local name. They will sometimes also exude a sticky white substance from their skin, which is presumably distasteful to predators.

Pale Frog *Litoria pallida* 1½ in | 4 cm

This is a small frog with a pointed snout and very long legs. It is brownish above and, although sometimes plain, usually has darker blotches or mottling. It has a black line through the eye, and the males have a yellowish upper lip when breeding. It is quite common throughout the Top End and is always found close to water, with large numbers gathering around the edges of swamps, wetlands and even small ditches or drains. A common call is a downward-inflected "*rrrt*", repeated once every second, but it often gives a variety of other shorter notes as well.

Where to find Grassy areas near water just about anywhere.

Peter's Frog *Litoria inermis* 1½ in | 4 cm

This species is very similar to the Pale Frog in size, shape, colouration and vocalization, and the two can be difficult to tell apart, except for the experienced herpetologist (frog and reptile expert). However, the back of Peter's Frog is covered in distinctive small warty lumps (it is smooth in Pale Frog). The two species are often found together, occurring throughout the Top End near water in grassy habitats including savannas and floodplains. Both species have relatively long legs, which they use to great effect when disturbed, as they are able to leap great distances. If they hop away in long grass, they travel so far they are almost impossible to refind.

Where to find Grassy areas near water.

Cane Toad *Rhinella marina* 8 in | 20 cm

Cane Toads are native to Central and South America, but were introduced to Australia in 1935 in an attempt to control two species of native beetle that were affecting sugar cane crops in northeastern Queensland. First released near Cairns, they spread down the east coast as far as northern New South Wales, and across northern Australia, reaching Kakadu NP in 2001, and recently the Western Australia border. The toads defend themselves against predators with a toxin, lethal to most Australian animals, that they release from the large parotid gland behind the eye. Many predators, such as snakes and goannas, that naturally feed on frogs suffer severe population declines after the Cane Toad arrives in an area, with population declines in some species of goanna reported at more than 90%. It seems that over time most species learn not to eat the toads and their populations recover, but this can be a slow process. Puzzlingly, there are different effects on species in different locations. In northeastern Queensland, populations of Northern Quoll (*page 200*) seem unaffected by Cane Toads, perhaps because they have coexisted now for nearly 80 years, but quoll populations in Kakadu NP declined severely after the toad's arrival. These differences have led to some ecologists questioning whether the Cane Toad is indeed a significant pest. Some animals are able to eat Cane Toads without any ill-effects, including the Keelback (*page 252*), while Torresian Crows (*page 186*) and Black Kites (*page 88*) have learned to avoid being poisoned by flipping the toad over and eating just the belly.

Where to find Very common and easily found at night around towns and cities.

Further reading

A book this size can only hope to cover a small proportion of all the wildlife that is found in the Top End. We have tried to focus on the species that you are most likely to see in your travels, or which only occur in the Top End – but if you find something that you cannot identify from the pages of this book, do not despair. There are a number of very comprehensive works that cover all of the birds, mammals, reptiles and frogs found in Australia, and the list below contains a selection of the best.

Birds

Birds of Australia: Eighth Edition (2012). K. Simpson and N. Day. Princeton University Press.

Birds of Australia: A Photographic Guide (2014). I. Campbell, S. Woods, N. Leseberg, with photography by G. Jones. Princeton University Press.

Finding Birds in Darwin, Kakadu and the Top End (2006). N. McCrie and J. Watson. NT Birding (out of print).

Finding Australian Birds: A Field Guide to Birding Locations (2014). T. Dolby and R. Clarke. CSIRO Publishing.

The Complete Guide to Finding the Birds of Australia: Second Edition (2011). R. Thomas, S. Thomas, D. Andrew and A. McBride. CSIRO Publishing.

Mammals

A Field Guide to the Mammals of Australia: Third Edition (2010). P. Menkhorst and F. Knight. Oxford University Press.

Field Companion to the Mammals of Australia (2013). Edited by S. van Dyck, I. Gynther and A. Baker. New Holland.

Tracks, Scats and Other Traces – A Field Guide to Australian Mammals: Revised Edition (2004). B. Triggs. Oxford University Press.

Reptiles

A Complete Guide to Reptiles of Australia (2013). S. Wilson and G. Swan. New Holland.

Frogs

Field Guide to the Frogs of Australia: Revised Edition (2011). M. J. Tyler and F. Knight. CSIRO Publishing.

Acknowledgements

We cannot acknowledge everyone who contributes to a book like this, but certainly the largest debt of gratitude is owed to the photographers who have provided all of the wonderful images. It is impossible to overestimate the hours required to take these stunning photographs. Days, weeks and often months are spent in in the field, searching for, finding and, with luck, photographing these difficult subjects. It is a labour of love that too often goes unrewarded, and we are very grateful to everyone who has provided photos for this book. We also owe special thanks to the **WILD***Guides* team: our editors, Andy and Gill Swash, designer Rob Still and proof reader Brian Clews. It is through their hard work that a chaotic combination of manuscript and photos has morphed into a readable volume. Thank you Andy, Gill, Rob and Brian ...

Photo credits

Every photograph published in this book, with the exception of the 127 taken by the author Nick Leseberg, is credited below, using the photographer's initials as follows:

Martin Armstrong [MA]; Nick Athanas (Tropical Birding) [NA]; Australian Wildlife/Creative Commons [AW/CC]; Michael Barritt [MB]; Julie Broken-Brow [JBB]; Simon Buckell [SB]; Iain Campbell [IC]; Roger and Liz Charlwood (WorldWildlifeImages.com) [R&LC]; Greg and Yvonne Dean (WorldWildlifeImages.com) [G&YD]; Frankzed/Creative Commons [F/CC]; Arthur Grosset (arthurgrosset.com) [AG]; Don Hadden (donhadden.com) [DHa]; David Hosking/FLPA [DHo]; Rob Hutchinson (Birdtour Asia) [RH]; Geoff Jones (barraimaging.com.au) [GJ]; Adam Scott Kennedy (rawnaturephoto.com) [ASK]; Deane Lewis [DL]; Andy Li/Creative Commons [AL/CC]; Katrin Lowe [KL]; Stewart McDonald (ugmedia.com.au) [SM]; Jane Menzies [JM]; Mike Menzies [MM]; Denzil Morgan [DM]; Pete Morris (Birdquest) [PM]; Terry Reardon [TR]; Jeremy Ringma [JR]; Laurie Ross (Laurieross.com.au) [LR]; Chris Sanderson [CS]; Andy and Gill Swash (WorldWildlifeImages.com) [A&GS]; David Tipling (www.davidtipling.com) [DT] Tom Tarrant (www.aviceda.org) [TT]; Martijn Verdoes/Agami [MV]; Scott Watson (Tropical Birding) [SWa]; Mike Watson (Birdquest) [MWa]; Martin B. Withers/FLPA [MWi]; Sam Woods (Tropical Birding) [SWo] and Stephen Zozaya [SZ].

Cover: **Gouldian Finch** [LR].
1 **Frilled Lizard** [SM].
9 **Varied Lorikeet** [LR].
20 **Star Finches** [PM].
21 **Torresian Imperial-pigeon** [SWa].
24 **Pied Heron** [G&YD]; **Comb-crested Jacana** [IC].
25 **Silver Gull**; **Black-fronted Dotterel** [A&GS]; **Little Kingfisher**; **Plumed Whistling-duck** [IC].
27 **Magpie Goose**: flight [AG].
29 **Wandering Whistling-ducks**: MAIN IMAGE [IC].
30 **Radjah Shelduck**: flight [PM].
31 **Australasian Grebe** [A&GS].
33 **Hardhead** [G&YD]; **Grey Teal** [TT].
34 **Black-necked Stork** [KL].
36 **Australian Pied Cormorant** [TT]; **Little Pied Cormorant** [A&GS].
39 **Australian Pelican**: flight [A&GS].
40 **Cattle Egret** [AV], flight [A&GS]; **Little Egret**: flight [A&GS]; **Plumed Egret**: [F/CC] flickr.com/photos/frankzed/12621621203, flight [R&LC].
41 **Great Egret**: [IC], flight [A&GS]; **Little Egret** [IC].
42 **White-faced Heron** [A&GS].
43 **Pacific Reef-heron**: dark morph [TT], white morph [DHa].
45 **Black Bittern** [A&GS].
46 **Straw-necked Ibis** [SWa].
47 **Royal Spoonbill** [DL]; **Glossy Ibis** [DL].
48 **Chestnut Rail** [LR].
49 **Buff-banded Rail** [A&GS]; **White-browed Crake** [IC].
51 **Common Coot** [PM].
52 **Brolga** [G&YD].
54 **Sooty Oystercatcher**: BOTH IMAGES [G&YD].
55 **Black-winged Stilt** [A&GS].
56 **Masked Lapwing** [PM].
58 **Red-kneed Dotterel** [PM].
59 **Black-fronted Dotterel** [A&GS]; **Red-capped Plover** [F/CC] flickr.com/photos/frankzed/9400619094; **Ruddy Turnstone**: non-breeding [A&GS].
61 **Grey Plover**: [A&GS], flight [AL/CC] flickr.com/photos/andy_li/6029811380; **Pacific Golden Plover** [A&GS]; **Lesser Sandplover**: non-breeding [A&GS]. **Greater Sandplover** [DHa].
62 **Oriental Pratincole** [DM].
63 **Australian Pratincole** [IC]; **Oriental Plover** [LR]; **Little Curlew** [DHa].
64 **Whimbrel** [DHa]; **Far Eastern Curlew** [MM]; **Bar-tailed Godwit** [F/CC] flickr.com/photos/frankzed/8228408709.
65 **Whimbrel** [A&GS]; **Far Eastern Curlew** [IC]; **Bar-tailed Godwit** [IC].
67 **Common Greenshank** [A&GS]; **Red Knot**: non-breeding [DT]; **Great Knot**: non-breeding [LR], breeding [SB].
68 **Red-necked Stint** [F/CC] flickr.com/photos/frankzed/6479155091
69 **Curlew Sandpiper**: flight [F/CC] flickr.com/photos/frankzed/8228408709, in water [TT]; **Terek Sandpiper** [A&GS].
71 **Marsh Sandpiper** [A&GS]; **Common Sandpiper** [A&GS]; **Sharp-tailed Sandpiper** [A&GS].
72 **Silver Gull**: [IC], flight [A&GS].
73 **Caspian Tern**: [R&LC], flight [MM].
74 **Lesser Crested Tern**: flight [IC]; **Greater Crested Tern**: non-breeding [A&GS].
75 **Gull-billed Tern**: flight [DM].
76 **White-winged Tern**: flight (breeding) [A&GS], flight (non-breeding), on ground [ASK]; **Whiskered Tern**: flight (non-breeding) [ASK], on ground [A&GS].
77 **Whiskered Tern** [A&GS]; **Little Tern** [A&GS].
78 **Azure Kingfisher** [IC].
79 **Little Kingfisher** [IC].
80 **White-bellied Sea-eagle** [MM]; **Black Kites** [IC].
81 **Wedge-tailed Eagle** [IC].
82 **Osprey**: [JM], flight [SWo].
83 **Brahminy Kite**: flight [MM], perched – BOTH IMAGES [LR].
84 **White-bellied Sea-eagle**: flight (adult), flight (immature) [A&GS], perched [MM].
85 **Pacific Baza**: perched [DL].
86 **Wedge-tailed Eagle**: BOTH IMAGES [A&GS].
87 **Black-breasted Buzzard**: BOTH IMAGES [DHa].
88 **Black Kite**: [NA], flight [IC].
89 **Whistling Kite**: juvenile [G&YD].
90 **Brown Goshawk** [IC].
91 **Grey Goshawk** [TT]; **Red Goshawk** [PM].
92 **Peregrine Falcon**: [GJ].
93 **Brown Falcon** [A&GS]; **Peregrine Falcon** [A&GS]; **Australian Hobby**: perched [PM], flight [DHa]; **Nankeen Kestrel**: perched [TT].
94 **Rose-crowned Fruit-dove** [LR].
95 **Rainbow Pitta** [CS].
96 **Orange-footed Scrubfowl** [SWa].

97 **Orange-footed Scrubfowl** [IC].
98 **Brown-capped Emerald Dove** [G&YD].
100 **Black-banded Fruit-dove** [NA].
101 **Rose-crowned Fruit-dove** [LR].
102 **Brush Cuckoo** [TT].
103 **Little Bronze-cuckoo** [TT].
104 **Rufous Owl**: adult [LR], immature [IC].
105 **Large-tailed Nightjar** [CS].
107 **Rainbow Pitta** [DHa].
108 **Collared Kingfisher** [SWo].
109 **Australian Yellow White-eye** [DL]; **Red-headed Honeyeater** [TT].
110 **Green-backed Gerygone** [TT]; **Large-billed Gerygone** [DHa].
111 **Varied Triller**: BOTH IMAGES [IC].
112 **Mangrove Golden Whistler**: female [TT], male [IC].
113 **Grey Whistler** [TT]; **Little Shrike-thrush** [JR].
114 **Australasian Figbird**: female [SWa], male [IC]; **Green Oriole** [KL].
115 **Northern Fantail** [TT]; **Arafura Fantail** [TT].
116 **Spangled Drongo** [TT]; **Black Butcherbird** [MWa].
117 **Broad-billed Flycatcher** [IC]; **Shining Flycatcher**: male [CS], female [AG].
118 **Buff-sided Robin** [RH].
120 **Variegated Fairywren** [G&YD]; **Gouldian Finch** [G&YD].
126 **Pheasant Coucal**: BOTH IMAGES [TT].
127 **Brown Quail** [DHa].
128 **Australian Koel**: male [LR], female [TT].
129 **Channel-billed Cuckoo** [A&GS].
130 **Crested Pigeon** [A&GS].
131 **Spinifex Pigeon** [LR]; **Common Bronzewing** [A&GS].
132 **Chestnut-quilled Rock-pigeon** [IC].
133 **Partridge Pigeon** [RH].
135 **Bar-shouldered Dove** [A&GS]; **Peaceful Dove** [A&GS].
137 **Southern Boobook** [IC].
142 **Blue-winged Kookaburra**: female [DHa], male [MWa].
143 **Forest Kingfisher** [IC]; **Sacred Kingfisher** [G&YD].
144 **Rainbow Bee-eater** [G&YD].
145 **Oriental Dollarbird** [A&GS].
146 **Red-tailed Black-cockatoo**: pair [IC].
147 **Cockatiel** [A&GS].
148 **Sulphur-crested Cockatoo**; **Little Corella**: ALL IMAGES [A&GS].
149 **Galah**: ALL IMAGES [A&GS].
150 **Red-collared Lorikeet** [G&YD].
151 **Varied Lorikeet** [LR].
153 **Red-winged Parrot** [LR].
154 **Hooded Parrot**: BOTH IMAGES [DHa].
155 **Hooded Parrot** [IC].
156 **Great Bowerbird** [IC].
157 **Great Bowerbird**: at bower [DHa].
158 **Variegated Fairywren**: male [G&YD], female [DHa].
159 **Purple-crowned Fairywren** [PM]; **Red-backed Fairywren** [IC].
161 **White-throated Grasswren**: BOTH IMAGES [RH].
163 **Banded Honeyeater** [LR]; **Brown Honeyeater** [IC].
164 **Yellow-tinted Honeyeater** [LR].
165 **Rufous-throated Honeyeater** [IC]; **Rufous-banded Honeyeater** [TT]; **White-gaped Honeyeater** [SZ]
166 **Bar-breasted Honeyeater** [LR]; **White-lined Honeyeater** [IC].
167 **White-throated Honeyeater** [LR]; **Black-chinned Honeyeater** [PM].
168 **Little Friarbird** [IC]; **Blue-faced Honeyeater** [IC].
169 **Helmeted Friarbird** [AG].
170 **Yellow-throated Miner** [A&GS].
171 **Weebill** [DHa]; **Striated Pardalote** [KL].
172 **Grey-crowned Babbler** [DHa].
173 **White-breasted Woodswallow** [IC]; **Black-faced Woodswallow** [IC]; **Little Woodswallow** [DL].
74 **White-winged Triller**: breeding male [G&YD]; **White-throated Gerygone** [TT].
175 **White-bellied Cuckooshrike** [TT]; **Black-faced Cuckooshrike** [G&YD].
176 **Varied Sitella** [LR]; **Black-tailed Treecreeper** [LR].
177 **Crested Shrike-tit** [RH].
178 **Rufous Whistler**: female [G&YD], male [IC].
179 **Sandstone Shrike-thrush** [DL]; **Grey Shrike-thrush** [SWo].
180 **Olive-backed Oriole** [G&YD].
181 **Pied Butcherbird** [A&GS].
182 **Willie-wagtail** [A&GS].
183 **Magpie-lark**: female [AW/CC] flickr.com/photos/47716821@N03/4372849790, male [G&YD].
184 **Paperbark Flycatcher** [DHa]; **Leaden Flycatcher**: male [TT].
185 **Lemon-bellied Flycatcher** [IC]; **Jacky Winter** [G&YD].
186 **Torresian Crow** [DHa].
187 **Australasian Bushlark** [DL]; **Golden-headed Cisticola**: non-breeding [JR], breeding [G&YD].
188 **Mistletoebird**: [A&GS], INSET [G&YD].
189 **Fairy Martin**[A&GS]; **Tree Martin** [A&GS].
190 **Crimson Finch** [LR].
191 **Star Finch** [LR].
192 **Gouldian Finch**: red-headed [LR], black-headed [G&YD]; **Long-tailed Finch** [LR]; **Masked Finch** [LR].
195 **Chestnut-breasted Mannikin** [G&YD]; **Yellow-rumped Mannikin** [DHa]; **Pictorella Mannikin** [LR].
196 **Agile Wallaby** [IC].
197 **Black-footed Tree Rat** [MA]; **Black Flying-fox** [IC].
199 **Short-beaked Echidna** [JBB].
200 **Northern Quoll** [MA].
201 **Northern Brush-tailed Phascogale** [MA].
203 **Common Brushtail Possum** [MWa].
204 **Rock Ringtail Possum** [MB].
205 **Sugar Glider** [DL].
206 **Agile Wallaby** [SWa].
207 **Northern Nailtail Wallaby** [MA].
210 **Black Wallaroo**: male [MA].
211 **Wilkin's Rock-wallaby** [KL].
214 **Yellow-bellied Sheath-tailed Bat** [TR].
215 **Yellow-bellied Sheath-tailed Bat** [SM]; **Orange Leaf-nosed Bat** [SM]; **Ghost Bat** [DL].
216 **Common Rock Rat** [SM].
217 **Black-footed Tree Rat** [MA]; **Water Rat** [JR].
218 **Dingo** [JR].
219 **Water Buffalo** [DHo].
220 **Green Tree Snake** [SM].
225 **Northern Yellow-faced Turtle** [IC]; **Northern Long-necked Turtle** [JR].
226 **Pig-nosed Turtle** [MWi]; **Northern Snapping Turtle** [SZ].
230 **Northern Spiny-tailed Gecko** [SZ].
232 **Giant Cave Gecko** [JR].
234 **Top End Fire-tailed Skink** [SZ].
235 **Slender Rainbow-skink** [DL]; **Red-sided Rainbow Skink** [SM].
236 **Common Blue-tongue**: top [SZ]; bottom [SM].
237 **Swamplands Lashtail** [SWa].
238 **Frilled Lizard**: threat display [SZ].
241 **Children's Python** [SM].
243 **Water Python** [JBB].
245 **Oenpelli Python** [SM].
246 **Carpet Python** [SM].
247 **Brown Tree Snake** [SM].
248 **Green Tree Snake** [SM].
249 **Taipan** [SM].
251 **Northern Brown Snake** [SM]; **Northern Death Adder** [SZ]; **Mulga Snake** [SM].
252 **Keelback** [SZ].
253 **Arafura File Snake** [SZ].
256 **Northern Dwarf Tree Frog** [JR].
258 **Dahl's Aquatic Frog** [SM].

Index

Names in **bold** highlight the species that are afforded a full account.
Bold numbers indicate the page number of the main species account.
Italicized numbers relate to other pages on which a photograph appears.
Numbers in normal text indicate pages from which species with a full account are cross-referenced, or other species that are referred to in the text.